Boomer Soldier;
A Hot War – Cold War Story

Boomer Soldier;
A Hot War – Cold War Story

Walt Cross
Master Sergeant
U.S. Army Retired

DIRE WOLF BOOKS
Stillwater, Oklahoma
502 E. Liberty Avenue
Stillwater, Oklahoma 74075-2630

First Edition 2018

Copyright © 2018 by Walt Cross. All rights reserved. No part of this work may be reproduced or transmitted, in any form or by any means, electronic or mechanical, including photocopying, recording, or any information storage and retrieval system without permission in writing from the publisher.

ISBN-13: 978-0-9850996-9-5

Cover created by Walt Cross. Photograph of the S-1 Section of Company A, 237th Engineer Battalion in the winter of 1971 during a field training exercise (FTX) in the forests of Germany.

Cataloging Data

Cross, Walt
Boomer Soldier; A Hot War – Cold War Story.
1. United States History – Vietnam War. 2. Cold War, Europe. 3. United States Army History-20th Century.

MANUFACTURED IN THE UNITED STATES OF AMERICA

TABLE OF CONTENTS

Chapter I	High School Days	1
Chapter II	Army Enlistment, Vietnam Era	9
Chapter III	A North Korean Act of War	21
Chapter IV	Fort Benning, Georgia	29
Chapter V	Crossing the Pond	35
Chapter VI	Return to the Real World	73
Chapter VII	Overseas to Europe	87
Chapter VIII	Co A, 237th Engineer Battalion	97
Chapter IX	Friends, Human and Animal	112
Chapter X	Terrorism Rises in Germany	133
Chapter XI	U.S. Army Recruiting Command	145
Chapter XII	Garrison Duty – Fort Sheridan	175
Chapter XIII	1st Battalion, 291st Infantry Regt.	193
Chapter XIV	Fifth Medical Group (USAR)	211
Chapter XV	Return to the 1st Battalion	221
Appendix	A Vietnam Photo Album	233
About the Author		287

FOREWORD

My father, Earl Lee Cross, is a veteran of World War II and served as an armored cavalryman in the 87th Cavalry Reconnaissance Squadron of the U.S. 7th Armored Division. He fought in the *Battle of the Bulge*. He was my direct family member of the Greatest Generation.

Dad suffered after the war with what has become known as 'post-traumatic stress disorder' or PTSD. After the war he served in the 2nd Armored Division, 66th Armored Regiment at Fort Hood, Texas. While he was stationed there I was born in the Fort Hood Army Hospital on April 20, 1949.

After he left the service in September of 1949 dad spent most of his adult life as an automobile mechanic. He was quite satisfied with simply working on engines, tires, brakes, and the other mechanical devices on cars. He never wanted to be a boss or foreman a fact that I believe was a symptom of his PTSD.

For as long as I can remember, dad talked to me about his experiences in the Army. He had no qualms telling me what combat was like. And I knew at an early age that I too wanted to be a soldier.

I met my wife Carol in the summer of 1967 and we fell in love. With the Vietnam War going on at the time I knew I was likely to be drafted, so in October of that year I asked Carol to marry me and when she said yes I enlisted in the U.S. Army. At this writing we recently celebrated our fiftieth anniversary on December 24, 2017. Carol traveled the world with me and we got to visit some pretty exciting places especially Hawaii and Europe.

For these reasons and numerous others, I dedicate this book to my dad, and my wife Carol. They are the two most influential people in my life. It is further dedicated to the Boomer Generation and their contributions to the nation while marching behind the Greatest Generation.

> Walt Cross
> Master Sergeant
> U.S. Army, Retired
> Stillwater, Oklahoma

So I will continue to continue, to pretend
My life will never end
And flowers never bend
With the rainfall

> *-Simon and Garfunkel*

Prologue

The Problem with being Humble

Nearly every Retired Soldier I've encountered is humble. It's not in our character to promote ourselves. And therein lays a problem for the Army, your Army. Americans don't know who among them is retired from the military. Not only are we an extreme minority (only 0.6 percent of Americans are retired from the military), we are also a silent minority.

So here's the larger problem: Americans don't know their military. They live far from a military base where they can talk to Soldiers and they don't know who is the Guardsman or Reservist living among them in their own neighborhoods. Most Americans feel they should thank us for our service, but they don't really know why. They don't know what Soldiers do in their name to preserve and defend their way of life. They don't know the sacrifices or the cost to our own families.

And they'll never know if we don't tell them.

That's where you come in. There are now 970,000 Retired Soldiers living in towns and cities all across the nation.

So what should you do? Identify yourself to Americans. Wear the Soldier for Life (SFL) lapel button. Put the SFL window sticker on your car or in some other highly visible place. The SFL logo is a conversation starter. Continue to set the example for others by the way you live. Get involved with your local chapter of Veterans of Foreign Wars, Disabled American Veterans, American Legion, or any local organization that helps veterans and inspires Americans. Get involved in your neighborhood or town or county.

And when you do, be a little less humble. Don't be silent about who you represent. Tell them you're a Retired Soldier. Wear your Soldier for Life lapel button and proudly display your Soldier for Life window sticker. Tell them your Army story and why you served. Tell them what today's Soldiers are doing to preserve and defend their way of life. Inspire them. Don't be part of the silent minority. Help us connect Americans with [their] Army.

Mark E. Overberg
Lieutenant Colonel
U.S. Army Retired[1]

---

[1] February – May 2018 edition *Army Echoes* newsletter

Chapter I
High School Days

My sweetheart Carol and I, summer of 1967

In the fall of 1966 occurred one of those social events that can seemingly have an impact all out of proportion to what they should have. The foundation of this event, a problematic change at the time, was laid eight years in the past during the year of 1958.

The beginning year of this brewing problem was 1955, I was six years old, and just starting school in the city of Sun Valley, California.

That first school day my mom dressed me in khaki shorts and a brightly colored, button-up shirt and tennis shoes. She had my little brother and sister who were just four and two years old respectively, as well as my baby sister, to look after. So she wrote my name and address down on a slip of paper, gave it to me and said I should follow the other kids and go to school. So that is what I did.

It seemed a long way to go, much longer than it likely was. But I hung in there and finally arrived at a pleasant looking school building with a green lawn out front and rows of benches. I was soon a student of Sun Valley's elementary school. I don't recall either of my parents ever visiting the school.

As I approached the manicured lawn a girl, much older than I, took my written slip and addressing me by name, asked me if I were in kindergarten.

I had never heard this foreign word before and without knowing its meaning I blurted out "My mama said I'm in the first grade."

The girl nodded knowingly, took me to a particular bench, and sat me down. The rest of the day was spent getting enrolled although I didn't know that's what was going on. And when a grownup asked me a question I just handed the slip of paper to them and that seemed to answer whatever it was they needed to know.

I can hear you asking what does this have to do with the problematic event in 1966. Well, I spent the next three years in that Sun Valley grade school and enjoyed it very much. We did schoolwork of course but I mainly remember the drawing, the painting, the lunches, and the naps. California schools were a lot of fun. And then, we packed up and moved to Oklahoma. Or I should say we returned to Oklahoma because that was where we had lived before relocating to California.

We returned to the small town of Cushing in north central Oklahoma where both sides of the family had lived since 1917 when my grandparents Walt and Linnie Cross settled there.

I attended Highland Elementary one of the town's two grade schools joining the third grade class in the middle of the school year. It didn't take my teacher long to discover I could print my letters fine, but when she asked me to write in cursive I had no idea what she was talking about.

A parent-teacher meeting later I found myself back in the second grade learning cursive writing. The teacher had convinced my mom that I was too far behind the rest of the class in this (at the time) vital skill. This form of writing is nearly outmoded today, but my lack of it put me a year behind my actual class. Not a serious setback of course, until it was magnified in 1966.

By the fall of 1966 I was in high school in the small northern Oklahoma town of Tonkawa where we had lived for the past four years. There, I played football, baseball, and participated in track all four years, and I was a letterman. I had even spent a term as vice-president of the class with a smart and pretty girl named Kathy Allen who was the class president. I was completing my second football season on the senior varsity team playing on offense, defense, and special teams. I was looking forward to my graduation in 1968. It was then my dad told me we were going to move to the town of Perry, about thirty miles south of Tonkawa. It might as well have been on the far side of the moon. I was pretty shattered.

Perry was okay, in fact I had gone to school there in the sixth and seventh grade before moving to Tonkawa and played on the junior high football team. But Tonkawa was everything I wanted high school to be. I went to Perry, but my heart was not in it, I seemed to be mired down. I finished the school year there in the spring of 1967 and started back to school that fall, but I just couldn't get myself motivated and I was unhappy.

I developed a chip on my shoulder and I soon found myself crosswise with the superintendent of schools who had his office in our high school.

I was already eighteen and soon found myself suspended for driving my car off a recently closed, by order of the superintendent, school campus.

I came back having served my three days off, and reported to the superintendent. He told me that he was going to give me six swats with a paddle and then forget that any of this had happened.

I felt I was an adult, and I had no intention of taking any corporal punishment. I was a grown man and used to defending myself. So, I told him that if he came from behind his desk I'd give him six swats.

My days of high school were over and I was expelled. That afternoon the local Noble County draft board called me and wanted to know what I was going to do now that I was out of school. I thought for only a moment and then answered that I was going to join the Army. The problematic event, begun in 1955, had borne fruit. I was just too impatient and too old to stay in school any longer. After all there was a war going on.

My dad was a member of the Greatest Generation, that generation of Americans who, through their sacrifice and service during World War II, no doubt saved the world from fascist and imperial domination.

I grew up on dad's stories of the war he experienced during his service as an armored cavalryman in the light tank company of the 87th Cavalry (Reconnaissance) Squadron, of the 7th Armored Division. He rode with the cavalry throughout the Battle of the Bulge and fought in the heroic stand of the 7th Armored Division at St. Vith, Belgium. It was always in the back of my mind that I would follow dad into the Army when it came my time and it was obvious to me, my time had come.

What I haven't told you during all this is that I had met the girl of my future, I'd met Carol Sellers.

Carol was from Texas, she and her parents moved to Perry in 1966 and I'd met her there. I first spoke to her at the old drive-in theater north of Perry one spring night in 1967. Later, Carol went to work at the 89er restaurant on the newly constructed Interstate highway 35 that ran north and south from Perry all the way to Dallas in the south, and north up to somewhere in Minnesota and ended at the Canada border.

She was a waitress earning money over the summer to return to school at Oklahoma State University in the fall. It just so happened I worked in the 89er Conoco service station attached to, but separate from, the restaurant.

We got to know one another and she told me she was born in the small town of Temple, Texas. I took this as some kind of omen because I was born at Fort Hood scarcely fifteen miles from Temple. Both our birth certificates showed our county of birth as Bell County, Texas.

Carol's mom and dad, George and Odessal Stafford, came to Perry when George took a job as the vice-president of the Triton Insurance Company.

Besides Carol, they had a second daughter named Rhonda and Carol's mom was also pregnant with her third child, Carol's future brother Lance.

Carol and I, despite her parent's concerns, were soon engaged to be married. In the meantime I enlisted in the Army along with two of my high school friends, Rick Myers and David Lemon that I recruited to go along with me.

Perry High School, the Maroons. School colors are silver, and of course, Maroon.

Chapter II
Army Enlistment in the Vietnam War Era

In our small town the news that three city residents enlisted in the U.S. Army at the same time was big news. The article below appeared on the front page, above the fold, of the *Perry Daily Journal* on October 25, 1967. Two days later all three of us were in the Army. David and Rick both became military policemen and I became a medical specialist.

The three of us reported to the Armed Forces Entrance and Examination Station (AFEES) at Oklahoma City. From there I would take my first commercial flight aboard Southwest Airlines to Fort Polk in the Bayou State of Louisiana. I was impressed by the jet plane and I still remember the stewardesses all wore what was termed psychedelic - colored hose.

We flew in the jet plane to Dallas where we boarded an Army prop-driven plane for the last leg of our journey to Fort Polk.

BOOMER SOLDIER

Three Enlist In Army For Three Years

Three Perry youths have been sent to Fort Polk, La., for basic training after enlisting in the Army for three years of service.

The three are Richard Lee Myers, 19; David Wayne Lemon, 19; and Walter Lee Cross, 18. They were examined in Oklahoma City last Thursday and Friday, and were sent to Fort Polk over the weekend. Each enlisted under the Army's "choice-not-chance" program.

Myers is the son of Mr. and Mrs. Leo Myers, 1114 Fir avenue. He has enlisted in military police. After basic training, he will receive extensive training in law enforcement at Fort Gordon, Ga. He is a 1966 graduate of Perry high school.

Lemon, son of Mr. and Mrs. Joe M. Lemon, 1010 Market street, also enlisted in the military police and will go to Fort Gordon after basic training.

The Perry

74th Year — No. 260 Wed., Oct. 25, 1967 (AP)

'If You Would Avoid

Lemon   Myers   Cross

Lemon is a 1966 graduate of Perry high school.

Cross is the son of Mr. and Mrs. Earl L. Cross, route 2, Perry. He enlisted in the Army Medical Service Corps. Following basic training at Fort Polk, Cross will receive special training at Brooks Army Medical center, Fort Sam Houston, Texas. Cross attended Perry high school.

The three were enlisted by Sfc. F. W. Godbee, stationed in Stillwater, as an Army recruiter. Young men interested in the Army's "choice - not - chance" program may obtain details from Godbee each Tuesday morning at the Army advisor's office on the south side of the square. He also may be contacted at 102 East Sixth street in Stillwater or by collect telephone call to Stillwater, FR2-4794.

"I was born in Forty-nine,
A Cold War kid in McCarthy's time,
Stop them at the 38th Parallel,
Blast those Yellow Reds to Hell…"

-Billy Joel
Leningrad

Distinctive Unit Insignia (DUI) of the 87th Cavalry Reconnaissance Squadron (Mechanized) WWII.

Unlike many veterans, my dad, a tanker in Europe during the *Battle of the Bulge*, never tired of telling me about it. And I never tired of hearing about it. From my earliest childhood I dreamed about becoming a soldier one day. I never missed the chance to play 'Army' with my neighborhood friends. I joined the Boy Scouts initially because they wore khaki uniforms that reminded me of pictures I'd seen of soldiers.

I remember being amazed to find out that soldiers actually got paid! I thought they served simply because they were super patriots. I had a dream when I was very young that I was standing on a smoking battlefield in a line of soldiers, and a smartly dressed Army officer was pinning a medal to my chest, it was red, white, and blue.

Years later a couple of days after a tremendous fight in Vietnam, a smartly dressed Army officer stepped up to me and pinned a medal to my chest, it was red, white, and blue. It's no wonder that when my 5th grade teacher asked me what I wanted to be when I grew up, I replied "I want to be a soldier".

I spent twenty-one years in the U.S. Army and when I was ready to retire the Army had done away with the Army fatigue uniform and replaced it with the camouflaged battle dress uniform commonly referred to as BDUs. Army Jeeps, deemed obsolete, were replaced by a vehicle called a Humvee and steel pots were replaced by the WWII 'German' looking Kevlar helmet. Since they had replaced all the things I had initially known as 'army', I decided it was time they replaced me. And guess what? They did. And when 911 happened in September of 2001, I volunteered at the age of fifty-two, to return to active duty but was told it 'was going to be a young man's war'.

What's in a Number?

Yes, I enlisted into the Regular Army on October 20, in the year of 1967. My service number was RA (for regular army) followed by seven numbers. Other soldiers I met later in basic combat training had the same RA prefix or NG (National Guard) ER (Enlisted Reserve) or the drill-sergeant-despised US (United States) which meant the soldier was a draftee.

Each time we entered the mess hall, we had to sound off with our service number at the top of our lungs. "RA16051747 drill serrrgeannntt!" we bellowed out as fast as we could. Then, depending upon whether you were RA, US, or other, the drill sergeant would tell you the appropriate amount of pushups to do before you got to enter the hallowed mess hall and partake of the Army chow awaiting you.

The NG's and ER's always got more pushups because of their status as 'weekend warriors', likely destined to never see duty in Vietnam. After their training they would return home to serve in Army Reserve units. They would drill one weekend a month, spend two weeks a year on annual training, and they serve the total of a six year obligation.

But the 'US', a member of the Army of the United States, the draftees of Uncle Sam's Army, only had to serve two years as compared with the three, four, or even six years of a man enlisting into the Regular Army.

So, these were the various categories of soldiers in basic combat training. Oh, there was one more category, sure, it was nominally a Regular Army status, but it was only a two year enlistment!

How could this happen? We pondered. How could a man have an RA designation before his service number and only have to serve for two years?

How indeed. He was a volunteer for the draft. Like the US, he would get no guaranteed job or assignment. For that reason he would only have to serve two years. In the view of some of the men in training, these two year volunteers had the best of all possible worlds.

But, returning to pushups, that most beloved pastime of Army drill sergeants, when yelling out your service number, if you started with RA in front of it he would reply "All the way!" and you would enter the mess hall. If, however, you yelled out one of the others it was "Get down and give me twenty!" (pushups). I always listened to the guys who weren't RA's to see if they would lie and escape the pushups, but none of them ever did. I don't know if it was a point of pride, hard headedness, or they were afraid the drill sergeant knew who they all were. Whatever the reason they never lied, just took those extra pushups.

So, here I found myself in the autumn of 1967, a member of Delta-Five-Two (D Company, Fifth Battalion, 2nd Basic Combat Training Brigade) at Fort Polk, Louisiana.

Fort Polk, Louisiana

Letterhead of the stationary I used to write home.

Fort Polk was established in 1941 and named in honor of the Right Reverend Leonidas Polk, the first Episcopal Bishop of the Diocese of Louisiana and a Confederate general. The fort is located in west central Louisiana among lots and lots of sand and pine trees and in the minds of military trainers 'A most suitable training site'.

I loved the smell of the pine trees. A major training camp during World War II, it was closed after the war. Ironically it was reopened in 1951 during the Korean War and my dad's old unit, the 7th Armored Division was stationed there. When the Korean War ended it was closed again.

The famous 'Check Point Charlie" during the Berlin Crisis.

The Berlin Crisis prompted the post's reactivation in 1961, and Fort Polk became an infantry training center in 1962. Three years later, it was selected to conduct Vietnam-oriented advanced infantry training in an area of the post designated 'Tiger Land'.

When I came to the post in 1967 it was divided into two basic combat training areas, South Fort and

North Fort, and D-5-2 (Delta Five Two) was located at North Fort. The Army was in the middle of reducing the number of 'general orders' soldiers were required to memorize from eleven down to three. Naturally the dividing line was the north and south fort, and south fort soldiers were only required to memorize three general orders. Lucky for me I was where soldiers had to learn all eleven.

A few days before I joined my BCT company, I was a civilian (later when assigned to recruiting duty I would tell prospective recruits "I ran away from home when I found out my mother was a civilian") living in the small town of Perry, Oklahoma.

Rick, David, me, and a soldier we met, David Edgar, were all from Oklahoma. I knew David and Rick in Perry and when we enlisted together it was on the buddy program. The buddy program simply guaranteed we would attend basic combat training together. After that, we separated.

David and Rick did go on to advanced training together since they both chose the same specialty. I would see David Lemon for the last time in November of 1969 in Vietnam. We went to the PX on the base at Phuc Vihn[1] drank a couple of beers together and caught up on personal events.

David was killed in 1975 after returning home to Perry and although the authorities claimed it was suicide; his family was convinced he was murdered.

Delta Five-Two

---

[1] Pronounced 'fook vinn'.

After his discharge Rick joined the Oklahoma Highway Patrol and had a full career, eventually obtaining the elected office of the president of the Oklahoma Highway Patrol's Fraternal Order of Police.

Marching was a big part of the training at Fort Polk. Note in the picture above our company is carrying the M-14 rifle that fired a .762 caliber NATO round, a much larger and heavier bullet than the .223 fired in the M-16.

Basic training was followed by three months of additional advanced training for me at Fort Sam Houston, Texas where I learned the job of a medical corpsman and combat medic. It was during this latter period of training that the dangers of my new profession as a soldier would become crystal clear.

Tiger Land in 1967

Between basic and advanced training I returned to Perry and married Carol there in the Methodist Church on Christmas Eve, December 24, 1967.

Since Carol and I were native Texans it seemed appropriate that when I reported to my Texas assignment I took Carol with me. Carol had family in San Antonio where I was to report at Fort Sam Houston for training. Carol's oil geologist Uncle Ariel, her mom's brother, lived there with his wife June and their three daughters. During my training we managed to spend some pleasant weekends with them at their very nice home. Their three young daughters; Leslie, Valerie, and Vanessa were still in high school.

While visiting, Ariel always hired me on to do some yard work and paid me well, giving Carol and I a needed boost to our family budget. We rented a small apartment off base but Carol spent a lot of time alone there as I often had to remain on the post, sometimes even over the weekends.

Money was tight and Carol often had to eat soup and bologna sandwiches.

BOOMER SOLDIER

Chapter III
A North Korean Act of War

Ship's crest of the *USS Pueblo*

In January of 1968 I started training at Fort Sam Houston Texas to become an Army field medic. Carol and I had married less than a month before. The event that was to place all the soldiers at Fort Sam on alert, including myself began some months before.

In the mid nineteen sixties Lyndon Johnson was president of the United States and it was at this time that a secret Navy surveillance program began code named *Operation Clickbeetle*. The purpose of the program was to create and deploy small surveillance vessels filled to the gunnels with electronic eavesdropping equipment. These vessels were to serve in a similar manner as electronic 'fishing trawlers' served as intelligence gathering ships for the Soviet Union.

In 1966 a former U.S. Army freighter designated FP-344 was transferred to the Navy and commissioned the *USS Pueblo* with the designation of AGER-2 (Auxiliary General Environmental Research). The *Pueblo*'s sister ship the *USS Banner* was AGER-1. The small ship was to operate under cover as a hydrographic research ship. Besides its crew of officers and sailors two scientists were assigned to make the cover story more plausible.

The *Pueblo* was only 176 feet long, weighed in at a featherweight 850 tons, and had a top speed of a mere 13 knots. Armed with two .50 caliber machine guns and small arms for the crew, the ship was vulnerable to any war ship or a better armed boat. The *Pueblo* was placed in the hands of Navy Commander Lloyd M. Bucher and assigned a crew of 83 men including officers. After a short shakedown cruise the ship was deployed on January 5, 1968 into the Sea of Japan for the purpose of conducting electronic surveillance along the east coast of North Korea.

Captain Bucher exercised careful attention to insure the ship stayed outside the twelve mile limit of territorial waters claimed by the communist country. Despite their meticulous navigation, less than three weeks into their first deployment the *Pueblo* came under assault.

On January 23, 1968 a North Korean sub chaser simply designated *No. 35* and armed with cannon confronted the U.S. ship in international waters. The confrontation may have been exacerbated when a North Korean attempt to assassinate the South Korean president was foiled with the subsequent death of the communist commandos sent to carry out the mission. This event took place the day before *Pueblo's* encounter with the North Korean.

These events were unknown to Captain Bucher who might have been even more cautious had he known of the attack. The North Korean sub chaser approached the U.S. Navy vessel and trained her guns on the *Pueblo* simultaneously hoisting signal flags demanding it prepare for boarders. The crew of the small ship could easily discern armed soldiers aboard the *No. 35* who trained their weapons on the Americans. Captain Bucher ignored the signal flags and hoisted his own that read I AM IN INTERNATIONAL WATERS.

The sub chaser continued to circle the *Pueblo* menacingly and was soon joined by three and later a fourth North Korean PT[2] boat. Bucher had his signalman hoist flags that read I INTEND TO REMAIN IN THE AREA. But the *Pueblo* was now surrounded by five hostile vessels with guns and torpedo tubes aimed at her. One of the PT boats came alongside the sub chaser, armed soldiers transferred to her, and the PT boat began to approach the *Pueblo* with the obvious intent of boarding the American ship.

The captain now ordered *Pueblo* to proceed toward open waters at one-third speed to avoid colliding with the circling communist vessels. The signalman now hoisted the signal THANK YOU FOR YOUR CONSIDERATION – I AM LEAVING THE AREA.

---

[2] Patrol boat, torpedo.

But *No. 35* replied with the signal HEAVE TO OR I WILL FIRE and sped up, closing the distance to *Pueblo*.

The intent of the North Koreans was made abundantly clear when two MiG jets roared over, flying close above the Americans. With the MiGs making a second pass the four PT boats broke away and *No. 35* angled for a clear shot which it soon delivered. Intended as warning shots one of the rounds struck a radar mast and sprayed sailors, including Captain Bucher, with shrapnel. The cannon fire was joined with that of chattering machine guns from the PT boats, the bullets smacking against the ship's sides.

Although Bucher immediately ordered the destruction of classified documents and equipment, much of the sensitive material subsequently fell into the hands of the communists, compromising U.S. surveillance operations for years to come.

Under assault, the *Pueblo* radioed an SOS requesting assistance and advising its higher headquarters in Japan of the situation. It was this message and subsequent messages that caused a general alert to all U.S. armed forces. I woke up on Wednesday, January 24 exactly one month after Carol and I were wed, to the possibility of immediate deployment to Korea without any further training.

In formation that morning our NCO instructors informed us that, depending upon the situation, our training as medics could be reduced to a 'crash' course. For the moment however, training would proceed normally until President Johnson indicated what direction the country's military was to take.

The *USS Pueblo*.

For the most part however, the crisis devolved into negotiations for the release of the *Pueblo*'s crew. Faced with overwhelming enemy firepower, Commander Bucher, after the death of one sailor by cannon shells and the wounding of several others, followed *No. 35* into captivity.

Their imprisonment would prove to be as horrendous as that experienced by soldiers, sailors, and airmen captured during WWII or the Vietnam War.

Captain Bucher would face an intense inquiry and recrimination by the Navy as would his crew for years to come after their release eleven months later.[3]

The *USS Pueblo* is still held by North Korea despite the fact it remains an active commissioned warship of the United States Navy. Early in 2013 the *Pueblo* was towed to Pyongyang, the capital of North Korea, and is tied up in the Potong River. There, it is usded as a museum ship at the 'Victorious War Museum'. It is the only U.S. Navy ship still on the commissioned list being held captive.

In 2018 during President Donald Trump's administration, and during a rapprochement with North Korean leader Kim Jong Un, the U.S. requested the return of the remains of our soldiers killed during the Korean War. The North Koreans assented and a number of the remains in North Korea were returned. It is expected, assuming negotiations continue to bear fruit, that more remains will be returned.

---

[3] For an excellent source of more information on the *Pueblo* Affair I recommend *Lyndon Johnson, North Korea, and the Capture of the Spy Ship Pueblo; ACT OF WAR* by Jack Cheevers, 2013 Penguin Group New York, NY. ISBN: 978-0-451-46619-8

In my opinion, the return of the *USS Pueblo* should be a priority as well. Since Kim Jong Un has stated he would like a documented end of the Korean War, now would seem to be the time for them to repatriate the *Pueblo*.

The *U.S.S. Pueblo* was taken illegally by force in international waters by an act of both piracy and terrorism. Her crew were imprisoned without due process and cruelly tortured for nearly a year! Looking back, I cannot condone the weak reaction displayed by President Johnson's administration in this naked act of aggression, regardless of other concerns he may have had.

As for me, I finished my medical training and moved on to my first duty assignment in the U.S. Army. Ironically however, I would virtually cross paths with Captain Bucher and the rest of the crew of the *Pueblo* in December of 1968. We crisscrossed as they were flying across the Pacific to come home and I was flying nearly the same but reverse route to join the war in Vietnam.

Chapter IV
Fort Benning, Georgia

My enlistment in the Army took me right out of high school in a small Oklahoma town, and trained me as a medical corpsman. After a short visit to her parents, Carol came with me to my first assignment at Fort Benning, Georgia. There, we lived in a trailer park outside the base in the city of Columbus while I served as a field ambulance driver in the 690th Medical Company (Ambulance).

I enjoyed the duty, which mainly consisted of providing onsite training support for the many training schools on the post. These included basic combat training, officer candidate school, airborne training, Army ranger school and some Special Forces training among others.

I drove my World War II era Dodge field (tactical) ambulance to a particular range and parked beside a field phone attached to a telephone pole. The phone was used by me to speak to my unit and for the training range officer or NCO to call me in case of a need for medical treatment and or transport. The cooks prepared me and the other medics performing post-wide medical commitments a take-along box lunch. It normally consisted of an apple or orange, a ham or bologna sandwich, chips, and cookies and came in a white cardboard box. We brought our own drinks or a canteen of water. Since I made seventy-five dollars a month whether I needed it or not, I brought my water-filled canteen.

I excelled in both my duties and my new role as a husband and despite never having enough money, Carol and I enjoyed the assignment. We rented an off base trailer house and with Carol's small black and white TV we set up or first home. Promotions came, first to Private First Class and then to specialist four and my pay and housing allowance grew to a whopping $145 a month!

The manner in which I reached the lofty pay grade of an E-4 (same pay grade as a corporal) is kind of interesting. I came in to the unit motor pool early one morning after pulling an all-night medical commitment. I set about the usual duty of cleaning

BOOMER SOLDIER

I learned to drive a WWII era ambulance similar to this one in March of 1968. It had a standard shift transmission with four wheel drive and held up to four patients on litters.

In the fall of 1968 the 690th Medical Company (Ambulance) received this new model of field ambulance, replacing all the old models of the WWII era.

my ambulance by washing off the dust from Fort Benning's dirt roads and anything I might have carried inside the cab on my boots.

Seemingly out of nowhere the company first sergeant and the captain appeared. I saluted and reported to the captain, who proceeded along with the first sergeant, to grill me on medical procedures, post regulations, current affairs in the news, and ambulance operations. When they left me to finish up the cleaning, they both seemed pleased. What I didn't realize was that the two of them were holding an impromptu promotion board.

The following week I was called up during morning formation and promoted to specialist four. I was not expecting it this soon as I did not yet have the prerequisite 90 days in grade as a private first class (PFC). I later learned the captain had waived that particular requirement. I was pretty happy and so was Carol, it meant a significant boost in pay.

My early rank insignia consisted of, private first class (PFC), specialist four (SP4) and specialist five (SP5). The first two pay grades of private (E1) and private two (E2) had no insignia at the time. Later, private first class became a stripe a rocker stripe underneath and the single stripe went to the private E2. The specialist five was the same pay grade (E5) as a sergeant and I became eligible to use the base NCO club. I sewed my PFC stripes on, but by the time I made SP4 Carol had joined me at Fort Benning and she sewed my rank on and starched my uniforms by hand in the bathtub and then ironed them for me. She did a great job and accepted her duties as a soldier's wife without complaint.

On April 4, two weeks before my nineteenth birthday, Martin Luther King Jr. was assassinated. Once again I found myself on alert and I spent several days and nights in my ambulance awaiting a possible call to Washington D.C. to help in quelling the race riots ongoing there. The seriousness of the situation was underscored when we were issued shoulder holster Colt .45 automatic pistols. Soon however, except in the arena of politics, the immediate situation calmed down.

Things continued in this happy manner until late in the summer of 1968. It was then that I and sixteen other 690th Medical Company medics were alerted for overseas duty to Vietnam. Carol and I went home on a thirty day leave and in December I flew off to Vietnam.

Before I left for Vietnam I visited my family in Oklahoma. Left to right is my sisters Kim and Theresa, and my younger brother Robin who is trying on some of my uniforms. Robin later joined the Army and became a helicopter mechanic and crew chief and became a sergeant. He served overseas in Korea. The dog is Lady, the smartest dog I ever knew. She seemed to know we were taking a picture and she posed with us. December, 1968.

Enjoying the sun in Vietnam at Fire Support Base Jim in 1969.

Chapter V
Crossing the Pond

My war in Vietnam began in December 1968 and lasted until January of 1970 with the entire year of 1969 sandwiched in between.

On January 30, 1968 the Viet Cong and North Vietnamese Army (NVA) launched an assault that became known as the 'Tet Offensive'. The liberal press, led by Walter Cronkite of CBS News, touted the offensive as the death knell of American efforts in Vietnam. In truth the offensive was a military defeat for the Communist North, but what they failed to achieve on the battlefield was won in the U.S. media and the minds of the American people. From that time on, Vietnam was seen by the majority of the American people as a lost cause.

I served in Alpha Battery, 1st Battalion, 7th Artillery of the 1st Infantry Division known as the Big Red One, earning four campaign (battle) stars on my Vietnam Service Medal. I also received two Bronze Star Medals, one for heroism in ground combat denoted by a "V" for valor device and an Oak Leaf Cluster denoting a second award of the same medal, this one for meritorious service in a combat zone. I also earned a promotion to specialist five, the same pay grade (E5) as a sergeant.

Rest and Recuperation (R&R) in Hawaii

I can give you the exact dates I spent in Hawaii with Carol, because the day I returned to Vietnam is the day Apollo Eleven landed on the moon. That date is July 21, 1969. I had spent the previous seven days in the paradise of Hawaii, made even more of a paradise when compared to Vietnam. That means that Carol and I began the greatest vacation of our life on July 14, 1969.

When my plane touched down and I, along with dozens of other soldiers, airmen, marines, and sailors got off, there was a long double line of wives and sweethearts waiting for us. I sauntered down the line in my khaki uniform all tanned a dark brown, my hair and mustache bleached almost blonde by the

Southeast Asian sun. When I walked up to Carol she was busy looking around me for me. She did not recognize me until I spoke to her. My first words were "Hey good lookin', who you lookin' for?" She gave me a startled look and then burst out "Walt!" It was one of the best homecomings I ever had.

We spent that week getting to know one another again, exploring the beaches and mountains of Hawaii, and eating steak and eggs every morning for breakfast. At least I did. But there was a dark lining to the vacation sun of Hawaii and I got a taste of what I would experience for years to come. I found I was very uncomfortable around groups of people and I was constantly looking around and on my guard. A car backfired while Carol and I were walking down a Honolulu street and I hit the pavement. The recognition of post-traumatic stress disorder or PTSD[4] was still years away, but I already had it.

---

[4] Post traumatic stress disorder (PTSD) is an anxiety disorder that can develop after exposure to one or more terrifying events in which grave physical harm occurred or was threatened. It is a severe and ongoing emotional reaction to an extreme psychological trauma.

On Monday July 21, 1969 Apollo Eleven landed on the moon and Neil Armstrong was the first man to walk on the lunar surface. It was also the day I got back to Vietnam from Hawaii so I got to watch the landing on a black and white TV at 1st Battalion headquarters before I returned to the field. I later got to see and meet Neil Armstrong in person at the Bob Hope Christmas Show that December at the 1st Infantry Division field headquarters in Lai Khe. I just happened to be at the right place at the right time to get to shake the hand of one of the most famous American astronauts. This blurb appeared in the *Stars and Stripes*:

Lai Khe, South Vietnam, December, 1969: *Apollo 11 astronaut Neil Armstrong, who became the first man to set foot on the moon earlier in the year, talks to U.S. troops during a surprise appearance with Bob Hope's Christmas show at the 1st Infantry*

Division headquarters. Armstrong was on a goodwill tour in Thailand when he was asked by Vice President Spiro Agnew to make a detour to Vietnam.

An offer came from Division leadership to allow anyone wishing to see the Bob Hope show at Lai Khe could catch a ride on the supply chopper that brought out our hot lunch each day. I was the only member of Battery A to accept the invitation.

So while the astronauts were flying to the moon, Carol and I were flying to Hawaii. She came from the U.S. mainland, and I from Vietnam by way of the small Pacific island of Guam.

I came home a twenty year old combat veteran on my way to making the Army my career. Before I left the 7th Artillery behind, I was offered a promotion to staff sergeant, a thirty day leave home, R&R[5] to Australia or back to Hawaii where I had met Carol for vacation in July of 1969, plus a tax free reenlistment bonus. But upon conferring with Carol via mail, we decided it was best if I just came home. Little did we know it would take me another four years to make staff sergeant.

---

[5] Rest and Recuperation.

The Fight at Fire Support Base Jim

Duty officer's log, Headquarters 1st Battalion, 7th Artillery 9 September, 1969 (declassified).

Entry number 4 at the early morning hour of 1:27am begins the dispassionate recording of the fight for Fire Support Jim with the entry "A 1/7th receiving incoming (82mm)."[6] The note that dusters spotted mortar tubes refers to a tracked vehicle armed with twin 40mm cannons. These vehicles were situated on the south side of the fire base and spotted the flash of mortar rounds as they exited the tube. Another entry at 0139 hours reads "A, 1/7th to fire killer jr. [sic] everywhere except the northwest. Taking RPGs[7] through the wire." A little further down at 0135 hours is the chilling entry "1 man missing an arm."

Other pertinent and interesting entries followed on the next page of the duty log.

"0143 [hours] A 1/7 says FDC blown away & probably could not fire a mission at this time.

0148 Dustoff[8] to arrive from NW in 10 min.

0152 A.O. [aerial observer] 50E due on station [above FSB Jim] 10 min.

0155 Req. E.O.M [request end of mission] on HE [high explosive rounds] from C 8/6 [Charlie Battery 8th Battalion, 6th Artillery] incoming stopped Hunter-Killer[9] to work around perimeter.

---

[6] 82 millimeter mortar, a high explosive round.
[7] Rocket propelled grenades.
[8] Helicopter to pick up wounded soldiers.
[9] An aerial Hunter-Killer team identifies and attacks enemy threats.

0155 30 [LTC Sperow, battalion commander] wants arty to the east of hwy through FSB Jim – Hunter-Killer to work western ½ [of the base].

0158 Dynamite A 8/6 [Alpha Battery 8th Battalion, 6th Artillery, code name "Dynamite"] up and ready [for fire missions].

0200 Nighthawk[10] [surveillance helicopter] in area will go take a look.

0210 Sit. Rep. [situation report] Sapper got through wire & hit FDC.

0217 A.O. 50E out of Phu Loi.

0218 A 1/7 to cut off Killer Jr. for Hunter-Killer team and Spooky[11] to work over area. [A 1/7] EXP [expended] 368 HE [high explosive] 140 Self Illumination [rounds].

This was a lot of support for us that arrived very quickly. The attack on Charlie Battery 1/7th four

---

[10] To deny the enemy freedom of movement at night a "night fighter" helicopter was developed and equipped with night sensors. The sensors consisted of a mounted "NOD" (night optical device, or starlight scope) and a Xenon searchlight with both infrared and white light capability. Armament was an M-134 multi-barreled minigun. Nighthawk was usually escorted by one or two gunships, and flew between 500 and 1,000 feet above the ground's surface.

[11] The Douglas AC-47 'Spooky' gunship (fixed wing, prop driven) carried massive firepower for engaging ground targets.

days before had served as a wakeup call for local command operations.

Killer junior is a reference to a deadly anti-personnel shrapnel artillery round. It was commonly called a "beehive" round and was packed with 8,000 flechettes, a French word for "little arrows". It is called a beehive round because of the distinctive buzzing noise the darts make flying through the air. This ammunition is devastating when used against ground troops due to its shotgun effect. When used effectively, it can sweep enemy soldiers from the battlefield, or nail them to jungle trees.

I was awakened around midnight by the sound of incoming mortar rounds. At first I thought it was the guns firing a mission, but the flat "crump" sound to the explosions alerted me to the fact it was something else. A 105mm howitzer has a distinctive "ring" when it sends a round down its steel tube. I woke up the other two guys in my bunker, one the battery commander's driver and the other a member of the "exec post" or headquarters crew. I told them to lock and load weapons and fire at the first thing that appeared in the doorway of our bunker. The VC was well known to send 'sappers', men armed with explosive satchel charges that move about tossing them into bunkers.

Suddenly, just above the sound of the explosions I heard "Doc, Doc, I got a man at the FDC (fire direction control center) with his arm blown off."

It was the battery commander, Captain Harry G. Madden. In my haste I tossed my .45 on my cot and grabbed my aid-bag. The only thing on my mind was what to do for a severed limb.

"Stay low, there's small arms fire and RPGs comin' in." The captain said as I exited the bunker.

I glanced to the left in the direction of the FDC[12] and saw numerous explosions in that area. A three quarter ton truck was parked midway between my bunker and the first of the gun emplacements with their sandbag walls. I low-crawled to the truck wearing nothing but pants and a helmet, my boots left behind, and took cover under it.

I could hear the stutter of AK 47s and the crack of rounds passing close overhead. It reminded me of basic combat training when we had low-crawled through strung barbed wire while an M-60 machine gun fired over our heads and quarter sticks of dynamite exploded around us.

Scanning the area ahead of me, I crawled to the first gun emplacement and looked back for the captain but couldn't see him. About that time the truck I had just left exploded as an RPG (rocket

---

[12] Fire Direction Center.

propelled grenade) struck it and ignited its fuel. It began to burn brightly and was completely destroyed.

To escape its revealing light I moved quickly through the warren of sand bagged gun positions, stopping behind each position before moving ahead. Laying on the ground and peering around a corner I was suddenly lifted from the ground, my ears ringing as a hand grenade exploded just on the other side of the sandbag wall I was hiding behind.

"I've been spotted!" I thought, and got to my feet, sprinting the last few yards and dodging around obstacles to the entrance of the FDC. As I arrived I saw him, a small VC soldier dressed in the usual black silk pajamas. He must have used up his last grenade, because he turned and ran. Seeing as how I didn't have a weapon, I was pretty glad he'd decided to do that. I stepped to the black opening of the FDC and yelled "Where's the wounded man?"

Receiving no answer I yelled it again. Still no one replied to my question. "Oh God, they're all dead." I thought, and moved into the bunker. The sharp stench of cordite, that telltale smell of a recent explosion, was so strong it burned my nose and throat. It is a distinctive smell that I have never forgotten.

I pulled my red lensed flashlight from my aid-bag and immediately saw a casualty at my feet. It was the soldier with the serious arm wound.[13] It wasn't completely severed and hung on by a narrow strip of flesh. I reached and pulled his arm into my lap.

A soldier named Mize, recently reassigned to us from the 82nd Airborne, held the flashlight for me as I applied a tourniquet. The wounded soldier, Don DeVore, who had just moved into the FDC the day before, lay on his back, his blood pumping out in long ropy strands to splash on my bare chest and arms. I adjusted the tourniquet, made from a medical cravat[14] and a stick.

When the blood flow finally slowed and then stopped I was relieved. I dressed the wound and then secured his arm to his side. Two of the soldiers in the FDC helped me place him on a medical litter. I wrapped a second cravat around his wounded arm and moistened it with water from a canteen.

I turned to Marvin Millhouse, the radio operator, and asked for a MEDEVAC dust-off chopper. While he radioed battalion headquarters I pulled chunks of shrapnel from his back and dressed his dozens of wounds until he collapsed from the constricting shredded muscles. Marvin later received the Army Commendation Medal with "V" for valor for

---

[13] Years later I learned this soldier was Donald E. DeVore when artilleryman Paul Jones saw DeVore give an interview on the History Channel in November, 2011.
[14] An olive drab colored scarf used for making splints for broken bones and tourniquets.

continuing to man the radio after being wounded. I treated Marvin for months to come, removing pieces of metal as they surfaced and fighting back against any onset of infection.

Sergeant Bob Abbott of the FDC picked up the PRC-29 radio and helped maintain contact with both Division Artillery and the dust-off chopper. The VC had managed to place an anti-tank mine atop the FDC and destroyed most of the radio antennas. Bob had interviewed DeVore that day to fill an opening on the FDC crew. He had told DeVore to move into the FDC that very night. When the fight began DeVore hunkered down behind a footlocker but left his arm lying atop the locker. When the first hand grenade went off it blew up a metal ammunition can that immediately turned into the shrapnel that severed DeVore's arm.

I treated a couple more soldier's wounds as the sounds of the continuing battle came to us. Our battery was up all six guns firing direct fire high explosives and the devastating 'beehive' round. These small but deadly missiles could nail a man to a tree, and often did.

Despite the incoming fire of mortars, RPGs and small arms, the dust-off was coming in. I organized a litter detail for the seriously wounded and led it out through the incoming fire and stopped them just inside the last sandbag parapet nearest the road running down the middle of the fire base, the only place the chopper could set down.

Low-crawling to the bullet-whipped road I pulled out a battery operated strobe light, and waited for the chopper. While lying on my back I watched the tracers arcing over our position and sometimes disappearing into the sandbag parapets. Illumination rounds catapulted parachute flares into the darkness to float slowly down, their swinging motion and red light causing shadows to come alive and dance. The occasional hot streak of fire denoted the passing of a rocket-propelled grenade followed shortly by a shattering explosion.

The enemy had no idea I was there, but they would in a minute. Hearing the chopper over the blast of the guns, I turned the very bright white-light strobe on and held it as high above my body as I could. The flashing light immediately drew enemy small arms fire and mortars began to fall around my position. But I could hear the chopper, and out of the darkness he came despite the tracers reaching for him and the fiery trail of an RPG as an enemy gunner tried to take the medivac down. He settled onto the road beside me, and as his door gunner fired into the enemy's position and our battery opened up with the hot exploding steel of protective fire,

I motioned the litter team forward then I and those brave soldiers loaded our wounded consisting of Don DeVore and Marvin Millhouse onto the chopper. I got the impression of a red haired burly warrant officer pilot and then with a roar and a cloud of dust the chopper lifted away, taking those men to safety. I returned to the gun sections and as they continued to fight the enemy throughout the night, I moved among them treating small wounds.

One individual I approached, a very young buck sergeant, was wounded in his left shoulder. He refused medical attention, continuing to lead his soldiers in the defense of our position. Seeing that it was only a flesh wound I didn't insist. Little did I know that my remark to the commander later about him refusing treatment would help earn him the coveted Silver Star medal for gallantry in action. We fought all night, and as the morning sun rose, the enemy melted away, taking his dead and his wounded with him.

In the heat and confusion of the assault, the crew of the number 3 gun just outside the FDC bunker, had thrown a thermal grenade down the gun's tube to keep it from falling into the hands of the enemy.

Some time passed and then a runner came from the engineer platoon on the north side of the fire support base. The engineers had just arrived at Jim that same night. He told me they had a man down and would I come take a look. I followed him to where the platoon sergeant lay unmoving on the floor of the engineer platoon's tent. I played my light over his face it was contorted as if in pain. I got no reaction from his pupils, they were fixed and dilated. I felt for a pulse, but there was none. I looked him over and could find no wounds. It was clear he was dead and I informed the engineers. I inquired later on with the Divarty[15] surgeon about the middle aged platoon sergeant and he told me he had died of an apparent heart attack. Had I been notified earlier I might have been able to save him.

The battery commander called me to the exec post and wanted the names of the evacuated casualties and those to be put in for the Purple Heart for wounds received in action. I filled out the report and the commander wanted to know why my name wasn't on the list for the Purple Heart and he pointed out the many cuts my feet and legs had sustained from my movements during the battle. I thought a moment about all the men who had sustained serious wounds and the engineer sergeant who'd been killed and then declined. I just didn't think my wounds warranted a Purple Heart. I told him about the sergeant who refused treatment and he made a note of it. He later recommended the artillery sergeant and myself be awarded the Silver Star for 'Gallantry in Action'.

---

[15] Division Artillery.

The Silver Star is the third highest decoration awarded for heroism. Needless to say I was pretty impressed at being recommended for it. But it was not to be. Back at Divarty HQ the recommendation was downgraded to a Bronze Star for Heroism (with a valor "V" device) for 'Heroism in ground combat', an impressive award.

The artillery sergeant who I mentioned to the battery commander as refusing medical treatment received the only Silver Star awarded to a 7th Artillery soldier in 1969.

The Story of Don DeVore

Many years later on Veterans Day, 2011 I received an email from 7th Artillery veteran Paul Jones. He told me that he had seen Don DeVore on the History Channel telling of his wound in Vietnam and that he was sure this was the soldier with the arm wound I treated. I found an article online about Don written by *New Jersey Herald* reporter Jessica L. Mickley and dated in early November, 2011.

A History Channel miniseries featuring [Don] DeVore and 12 others affected by the Vietnam War...premiered on Tuesday [November 8] ... "I'm flattered, but at the same time, it's a little overwhelming," DeVore said...Thirty-five years after serving in Vietnam, DeVore sought treatment for what he now knows is post-traumatic stress disorder. He attends a weekly group therapy session at Newton Medical Center with other veterans.

DeVore's tour of Vietnam lasted seven months. Midway through DeVore was sent home to be with his pregnant wife, who doctors thought might have complications during labor. He was in transit when his daughter was born, possibly somewhere over the Pacific Ocean, but he can't be sure. DeVore arrived on the third day of his daughter's life, just in time to usher his wife and baby girl home from the hospital.

DeVore had great timing, however, for the music event of a lifetime. He rounded up a few friends and went to Woodstock to see the Who.

"I think I was the only person there with short hair," he said. Returning to Vietnam afterward was "Dreadful," DeVore said. "That was the worst trip of my life."

Five weeks after landing in Vietnam, DeVore was severely injured by a rocket-propelled grenade [actually it was a hand grenade], *and spent the next two years of his life in the hospital. Scars from numerous surgeries remain as an unnecessary reminder.*

Nowadays DeVore is doing "All right." He does not love the attention resulting from his History Channel appearance (his children keep reading posts about him on Facebook). He's just one of many Vietnam veterans, he says, with many unique stories. DeVore participated in the documentary because he admired the program's intent in presenting Vietnam through personal stories. "It's not blood and guts," he said. "It's what people really went through."

In a related article the same reporter wrote the following also regarding Don DeVore.

Some veterans, like those who served in Vietnam, receive gratitude now, though that was not always the case. These veterans not only fought for their country and their lives, but came back to the states to fight an army of backlash.

"We were baby killers. We were all drug addicts," *Vietnam veteran Don DeVore of Franklin [New Jersey], said.*

DeVore, a regular attendee of the county Veterans Day celebration, was wounded by a rocket propelled grenade about seven months into his tour. He spent the next two years in the hospital, in and out of surgeries. He came out with an arm of scars and a lot of shame.

"There were times I wouldn't even acknowledge that I was a veteran," *DeVore said.*

For a while, when asked about his scars, DeVore would blame a motorcycle accident, until about 16 years ago. His then 16-year-old daughter entered an essay contest "What Democracy Means to Me", through the VFW.

"It's the first time it even really dawned on me that my children were really proud of my service," DeVore said.

The History Channel video DeVore appeared in is titled *Vietnam in HD* and the producer wrote a synopsis of DeVore's story.

Monticello, New York
U.S. Army, 1st Division. Service: Spring 1969 – Summer 1969.

Like thousands of other young American men, Don DeVore struggled intensely with what he would do if he were drafted to serve in Vietnam. He had no desire to become a war hero, and no dreams of winning glory or greatness on a battlefield.

In the late summer of 1968, DeVore's number was called and within weeks he was shipped off to basic training at Fort Jackson, South Carolina. Arriving in Vietnam in March of 1969, DeVore was assigned to an artillery unit at a fire support base known as Firebase Jim. His job was to provide accurate fire support for the search and destroy patrols that were taking place on a near daily basis in the surrounding jungles. After four months, DeVore was granted compassionate leave to attend the birth of his first child. Upon returning home, he found himself in the middle of the largest peace and love festival of the decade – Woodstock. It was a stark contrast to the harsh combat he returned to just days later. In September of 1969, the Viet Cong infiltrated Firebase Jim and DeVore was severely wound by an RPG [sic] rocket propelled grenade), sustaining an injury to his left arm that kept him hospitalized for nearly two years. The psychological and physical effects of his combat experience were devastating. For years, DeVore never spoke about the war. When questioned about the scars on his arm, he would tell people they were the result of a motorcycle accident. Finally, in the late 1990s, he sought treatment at a VA hospital, and after several years of counseling he was finally able to come to terms with [his] *wartime experience.*

In November of 2011 I found Don's address online and wrote him a short letter. A few days before Christmas, on December the 21st, I received an answer back from him. It had been 42 years since we last spoke to one another.

Doc Cross! Holy s**t. Man, it was great hearing from you! Before I go any further, please let me say "thank you" a thousand times for taking care of me! I know what you did – I read your book quite a while ago.[16] I wish you had written it sooner! I owe you brother! When I first heard about your book (and you, obviously), I Googled around and found you at OSU and tried to reach you. When I didn't hear back, I assumed you either were no longer at the school, or perhaps, you just weren't up to dredging up the past. Completely understandable! I know.

I am doing very well. I'm in good health, have a wonderful wife, six (6) great kids, and four (4) beautiful grandchildren. I've pursued a career as a professional engineer and land surveyor and currently work for the Base Ops contractor at a large, believe it or not, Army R&D installation near my home (Picatinny Arsenal, NJ).

Please, please, let's stay in touch. God bless you Walt. Thanks so much for reaching out to me! I'm looking forward to hearing back from you.

Don DeVore
Franklin, NJ 07416

---

[16] This is a reference to my book *Fastest Gun in the Big Red One*.

Don DeVore as he appeared in the History Channel Film.

The fight for Fire Support Base Jim was not yet over. Compared to the early morning hour's action, the second attack was not as intense and this paragraph from the unpublished history of the 7th Artillery sums it up as follows:

Evidently assuming the personnel at the fire support base dropped their guard because of the previous early morning activity, the enemy again attempted to breach the perimeter a second time that night. This time, however, the ground surveillance radar[17] set at the fire support base detected the activity and the enemy was engaged with small arms and "Killer Junior".

---

[17] Ground surveillance radar is ground-to-ground surveillance radar set for use by military units. The radar is capable of detecting and locating moving personnel and vehicles.

Heavy enemy blood trails indicated the enemy had suffered casualties again. No friendly forces were wounded during this second attack.

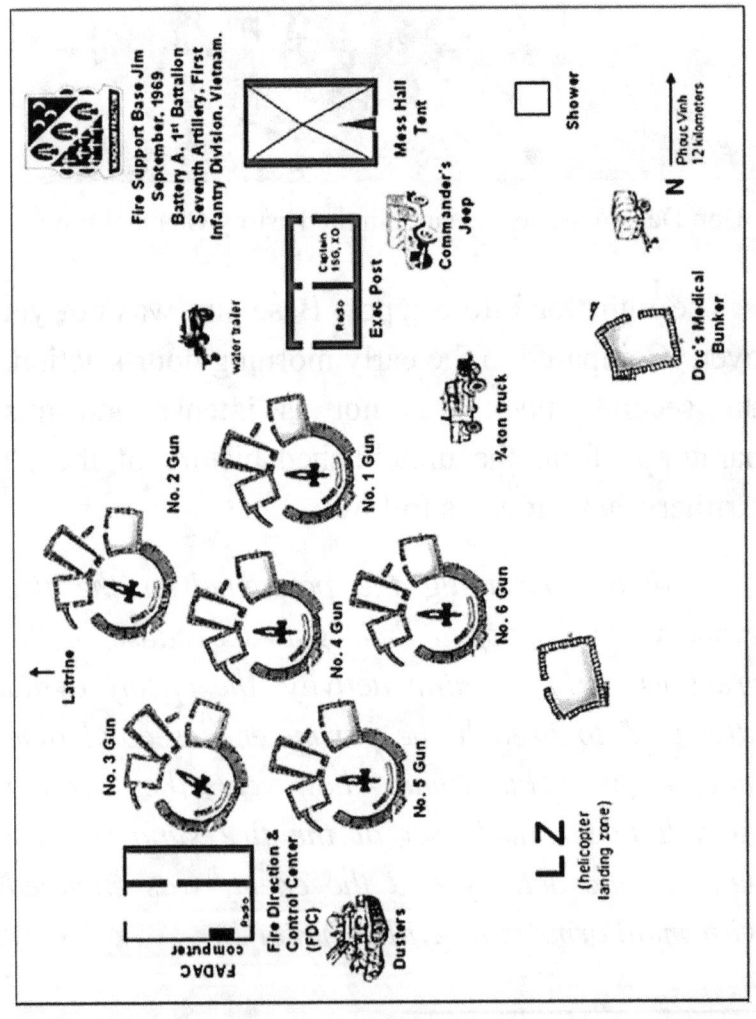

Diagram 1, Fire Support Base Jim September 8, 1969.

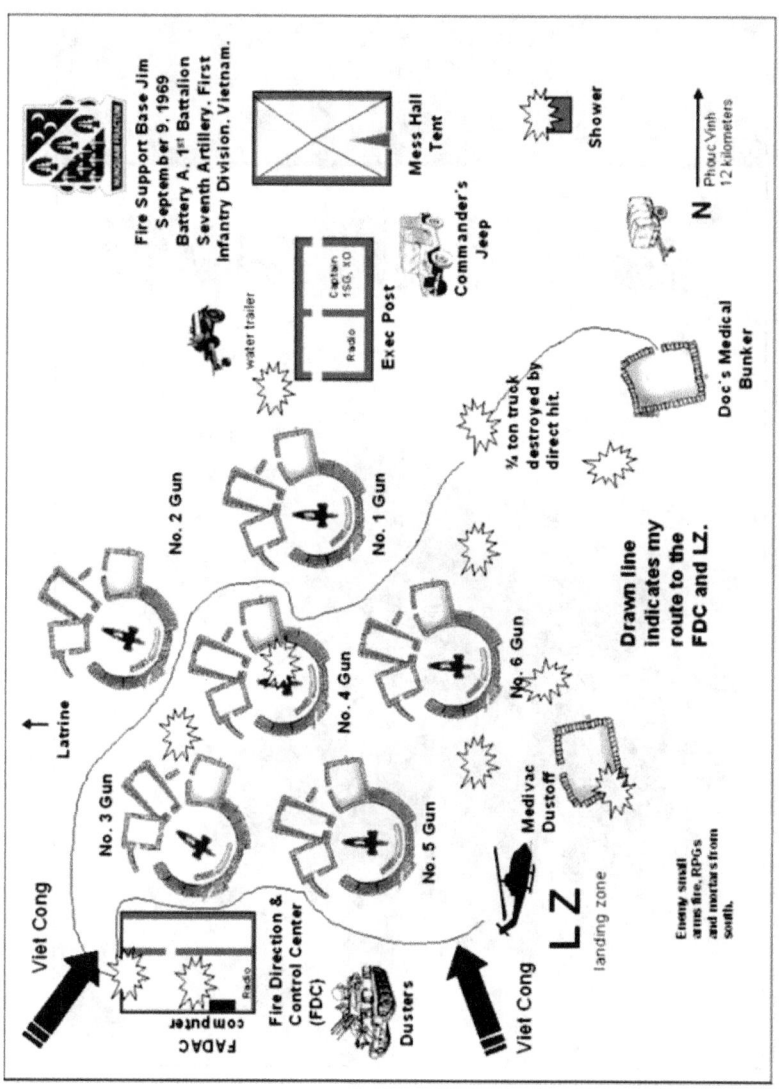

Diagram 2, Fire Support Base Jim September 9, 1969.

A Changing War 1969 – 1970

Three minutes into the second DVD appears the following title in white lettering: Fire Support Base Jim Spring 1969. The film footage is stock; the scenes depicted are not from the actual Fire Support Base Jim. Former Battery A, 1/7th Artilleryman Don Devore is introduced as an artilleryman of the First Infantry Division stationed at FSB Jim. There is no actual mention of Alpha Battery or the 1/7th Artillery. The narrator describes Devore's background and he is shown as he himself picks up the narrative. The film then goes on to other Vietnam vignettes, and returns to Devore a few minutes later.

Ten minutes into the documentary it returns to Donald Devore who continues to narrate his return home on emergency leave and attending the festival at Woodstock.[18] Don left Vietnam three days before the anti-Vietnam War festival. What a change of venue that must have been for him! Once again the title returns as Fire Support Base Jim and more stock footage of artillerymen is shown. Devore mentions the death of Ho Chi Minh on September 9, 1969 and the declared truce for that night. Don narrates the events of the attack as he remembers them. Don had lost a lot of blood and was in shock which likely colors his memories and his story when he states near the end of the film sequence that:

---

[18] The Woodstock Music & Art Fair, informally known as the Woodstock Festival or simply Woodstock, was a music festival, billed as An Aquarian Exposition: 3 Days of Peace & Music. It was held at Max Yasgur's 600 acre dairy farm in the Catskills near the hamlet of White Lake in the town of Bethel, New York, from August 15 to 18, 1969. It was anti-Vietnam War.

"When I opened my eyes a bunch of guys were cheering for me, kind of like a football player who has been hurt on the field now being carried off to the yells and the screams of the crowd. As we lift off [in the medevac chopper] a calm starts to come over me. I can see out the side door that the morning sun is starting to set in with a purpulish [sic] red sky. All I can think to myself is I'm alive! For me, this craziness is over."

With those words the sequence about Fire Support Base Jim ends. After pondering his words and impressions I can draw a few conclusions from them. I had not administered morphine to Don at the time he was wounded, a fact I imparted to the medic on the helicopter. Without a doubt the calm that came over him was from morphine given him by that medevac medic. As for the dawn statement, the helicopter arrived at 0148 hours, or a little before 2 a.m. in the morning. We were many hours away from dawn and his flight onboard the chopper was likely no more than twenty to thirty minutes in duration either to the division's forward operations base at Lai Khe, or to Saigon. Don's memory of a morning dawn is a false memory and may also be attributed to the morphine he received on the helicopter.

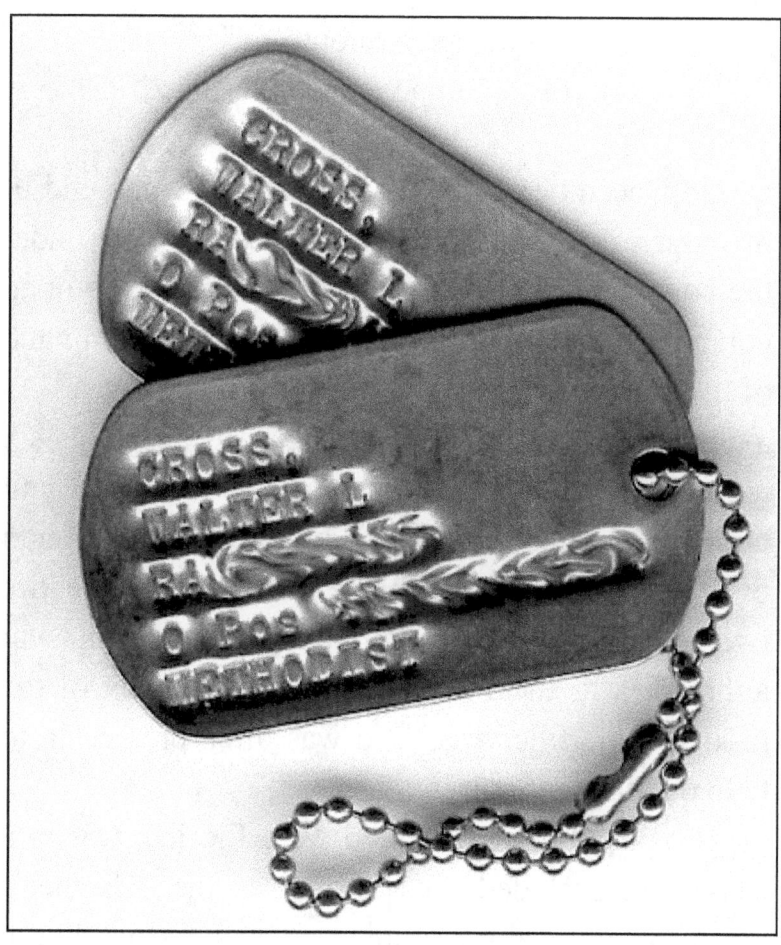

The dog tags I wore in Vietnam. Both my RA number and my SSAN were displayed on them.

"Enemy Helicopters[19]"
At Fire Support Base Florida

On December 29, 1969 I was nearing the end of my tour of duty in Vietnam The battery was under the command of Captain Harry G. Madden, a young courageous artillery officer who assumed command of the battery in July or August of that same year. As a battery commander's tenure was usually six months, Captain Madden remained with the unit until it was send home three months later in March of 1970. I had been with the battery for my entire twelve months of deployment in Vietnam, and although a medical aid-man, I learned many of the tasks of an artilleryman and was often pressed in to help out during fire missions.

The position we occupied, FSB Florida, was in a somewhat secluded area known as the "Catcher's Mitt" for its local topography of hills surrounding a valley. In the "mitt" part of the valley ran a section of the Ho Chi Minh Trail leading to the northern approaches to Saigon only about 30 kilometers south. The area was the operational domain of the 11th Armored Cavalry Regiment (11th ACR) as well as the Dong Nai Regiment of the Viet Cong. FSB Florida had no roads leading to or out of it. We were

---

[19] UFOs were not reported as such in the official records of the Vietnam War, but were always referred to as "Enemy Helicopters".

brought in by helicopter with our 105mm howitzers carried in one at a time by Jolly Green Giant transport choppers. While coming in the perimeter of the hill top position was secured by a heavy platoon of infantry. After we set up the guns and became operational the infantry would be taken out.

We had a good open view of the valley thanks to a defoliated jungle that reached out a good hundred yards around the fire base. Scattered here and there stood the naked trunks of trees we would use as target practice for our cannons in the coming days. This cleared the area even more, denying the VC or NVA any safe hiding places.[20]

A couple of nights into our stay I took a turn at perimeter guard from atop the exec post, a bunker that housed the headquarters personnel. The enemy often used flashlights to set up their mortars in the dark of the night, made even darker by the cover of the jungle canopy. To help spot these nighttime operations and bring our guns into play before the enemy could launch an attack, the battery had recently received a starlight scope.

The scope was excellent for picking up any ambient light source as well as amplifying any available ambient light, such as the faint light of the stars. As soon as it got dark I began to slowly scan

---

[20] The Viet Cong and North Vietnamese Army respectively.

the wood line seeking the telltale flash of an enemy light. But my eye was brought up when I caught the glow of something above the trees. I brought the scope up and there, south of our position and seemingly moving quite slow, appeared a flying disc-shaped object in the familiar green color projected by the starlight scope. It was flying level above the tree canopy and moving, relative to my position, right to left or west to east. I did not automatically think of it as a UFO, in fact I thought it must be one of our helicopters despite its distinctly disc-shaped fuselage.

    I visually tracked it with the scope as it turned in a wide arc and changed its direction of travel from east to north east and then north. Removing my eye from the scope I tried to make out the shape with my naked eye, but all I saw was the pitch black of the jungle. Although there were stars out and I watched for a shadow moving across them, I saw nothing. I also listened intently but could not detect the familiar "whump, whump, whump" sound of a helicopter.

    I returned my eye to the scope and it took me a moment to again gain a visual of the object. It had moved further north and as I continued to watch a second green lighted object appeared below the first. But this one was not flying and in fact was not a disc but a dome, and it sat on the southern slope of a hill easily seen in the daylight to our north east. What really caught my eye was the fact that the dome, although stationary, was pulsing slowly. The pulses seemed to come every five or six seconds, and I could see the disc tip its "nose" in relation to the dome and continue to move toward it at an angle, as though it were going to pierce the dome. I turned to one of the exec post personnel and said "Go get the captain, I got something." Then I returned to the scope and watched incredulously as the disc did indeed, pierce the dome and merged with it, slowly disappearing from view.

That is when Captain Madden showed up. "Whatcha got Doc?" he asked. I replied that I wasn't sure and invited him to take a turn with the tripod mounted scope.

"What the hell is that?" He continued. I told him what I had seen, the slow flying disc and the fact that it disappeared into the dome.

"Yeah, I see it. It appears to have launched from the dome, or maybe it's a different disc. It's heading away north and east." Then he lifted his view from the scope.

"What do you think sir?" I asked. The captain just shook his head.

"I don't know what to think. It doesn't appear to be a threat to us so ignore it and concentrate on watching for VC." He concluded and had started to walk away but then turned. "And I don't want any report on that appearing on the exec post log!"

I acknowledged his instructions and after looking at the dome for a while longer I went back to searching the tree line. After making a complete sweep of our perimeter I went back to where the dome was positioned, but it was gone as if it had never been there.

About thirty kilometers north of Fire Support Base Florida was the base camp of the 1st Cavalry Division. Among the soldiers stationed there was a hometown friend of mine, Specialist David Lemon. David and I grew up in the same small town in Oklahoma and enlisted in the Army together on the "Buddy" plan. David was a military policeman and worked on the main gate of the base. Around the time I spotted the "enemy helicopter" to the south and east of the base a perimeter guard at the base also saw an intruder, perhaps the same intruder I saw. The name of the witness was not given in the report it may have even been David. I did not find this report until many decades after I left Vietnam.

I never got to swap Vietnam stories with David, I was sent to Europe after leaving Southeast Asia while David went home, and died not long after. That report from Phuoc Vinh is presented below and offered as a supporting statement to the Fire Support Base Florida narrative. When last seen by the 1st

Cavalry Division soldier the intruder was moving south, in the general direction of Alpha Battery, 7th Artillery.

"Enemy Helicopter" Encounter at Phuoc Vinh North of FSB Florida

"I was pulling Green-line duty with three other 1st Cav. soldiers who were sleeping. I had a starlight scope, a radio and all the stuff you would expect in/on a bunker. This bunker was a big well-fortified bunker. We were all on top of this bunker from my best recollection. I was pulling my stint, letting the others sleep. This bunker was on the western facing perimeter. This night was a beautiful night with no overcast. Many small brilliant stars were in the night sky. No moon, as far as I remember especially facing out west. The starlight scope was working very well. I'm going into a little detail to set the stage leading up to my sighting. I know it's not to the point please bear with me.

As I was scanning the western night sky, all of a sudden something to the Northwest caught my eye. It was a very brilliant whitish, silver and with a hint of blue, more of a rounded shape. It was fairly far away. The main thing about this object was it would move to my left, or south in jerky movements, hover, do it again and again. It never lost the same brilliance or colors the entire time. Additionally it left an amber or reddish trail (like a tracer) as it moved only to suddenly stop on a dime. I watched this thing for several minutes.[21]

By this time it was in the southern horizon. Then, all of a sudden it shot up skyward on a 45 degree angle towards the north. It was totally out of sight in seconds. I was thinking should I report this? I decided not to. I would wait to see if anyone else would, then I would to. I didn't have the presence of mind to wake the other guys. Mainly, I couldn't believe what I was seeing."

While we were at FSB Florida the battalion tactical operations center moved to Lai Khe on December the 9th to support 2nd Brigade operations. The brigade was moving into a new area of operations north of Lai Khe between Thunder Road and the Song Be River located to the northeast of the big base camp. On December 12 Bravo Battery moved its six guns from FSB Jim to FSB Oklahoma in support of an operation by the 1st Battalion, 26th Infantry along the western edge of the Song Be River. I have no idea what happened to FSB Jim after the 7th left, but it was likely stripped of all

---

[21] He had to be viewing this object with the naked eye because a starlight scope shows all light images as green.

useable military materiel, deconstructed, and abandoned.

This first generation of starlight scope cost $20,000 when the battalion obtained them in April of 1969.

BOOMER SOLDIER

Chapter VI
Return to the Real World

I flew home, or as we all called it in Vietnam 'the real world' in January of 1970 aboard Flying Tiger Airlines. We landed at Oakland Army Airbase in California. There, Army contract tailors made me the best set of winter weight dress greens I ever wore during my service. They sewed my rank on, my 1st Infantry Division combat patch on my right shoulder, added new bright insignia on the colors and a full set of my military ribbons. I was and looked like a professional soldier. Then they warned me about the trouble I might expect if I wore my uniform off base and suggested I wear civilian clothes to the airport. I said to hell with that, I had just fought one war and I could fight another with the stupid protestors.

Then I went to the airport and had a big juicy hamburger and vanilla malt. They were delicious!

I picked up Carol from her parent's home in Houston, and we went to our next assignment at Fort Campbell, Kentucky. This post is the home base of the 101st Airborne Division. But the 101st was still overseas at that time and so there was plenty of post housing available and we moved into a two story apartment equipped with furniture and cooking utensils provided by the post quartermaster. Then we got a few other things we needed and became domesticated as outlined in these letters I wrote to my mom and dad.

SP/5 & Mrs. Walter L. Cross
4147A Lee Village
Fort Campbell, KY 42223

Mr. & Mrs. Earl L. Cross
Rt #1
Ripley, Oklahoma 74074

6 February 1970

Dear Mom, Dad, & Kids,

Well, I've got some good news. We're settled now [we had just arrived at the fort after my tour of duty in Vietnam]. We got some real nice quarters (housing). Two bedrooms upstairs (plenty of room for Rob[22] when he comes) or [for] all of you if you come this

---

[22] My brother, Robin.

summer. We bought a Spanish [style] couch corner set, a lamp that has a bulb in the bottom too, and something I've always wanted, a 20 inch color TV! How about that? I'm getting rich in this man's Army. The reason we could afford it was that I got $518.00 in travel pay, and we took a little out of the savings. Carol's drapes will be in, in a couple of weeks and when we get them up [I] will take a picture and send it to you all.

The couch is avocado and yellow gold. The lamp is a burnt gold and the drapes will be gold. But the TV is a portable, I couldn't quite swing $700.00 for a console.

Guess what? I'm not a medic anymore. I reported in yesterday, a guy [Staff Sergeant Cornelius] looked at my records and said wait a minute. A few minutes later, a major called me in [to his office] and asked if I wanted to work for him as a [personnel] clerk. I said yes, so now I spend my time beating a type writer and squaring away 201 files [military personnel files]. The hours are 7:30am to 4:30 in the afternoon. Just like an office job. The life of Riley. At least I don't have to hand[le] bed pans. Oh yeah, I get the weekend off too.

There's a place near here they call the 'Land Between the Lakes'. There are two large lakes with a peninsula between them. There are all kinds of swimming, boating, and camping places. And over towards Louisville is the 'Mammoth Caves'. So this summer there's going to be plenty to do. I even thought I'd go to Louisville and lay a couple of bucks on a horse. Come on and write. I know your hands aint broke.

                        Love,
                        Walt & Carol.

SP/5 & Mrs. Walter L. Cross
4147A Lee Village
Fort Campbell, KY 42223

Mr. & Mrs. Earl L. Cross
Rt #1
Ripley, Oklahoma 74074

12 March 1970

Dear Mom, Dad, and Kids,

Sorry I haven't written in so long, but I've been [so] busy that I don't know who I am.

Well, I did it. I probably shouldn't have but I did. I bought a 1970 Plymouth Duster[23] [my first new car]. It's a bright orange [competition orange] with black interior, power brakes, steering, and automatic transmission [what I didn't mention is that it lacked a/c]. We really like it. It's real sporty looking. I went down the other day and took my last college test. If I pass it I'll have a year of college [credit].[24] I also tried to get into the military police academy, but they said my eyes were too bad [actually I did get to go the academy at Fort Campbell and later became a CID[25] agent in Germany.] So there go my thoughts of being a cop. Now I've got to think of something to do. I get out [of the Army] in seven

---

[23] I owned this car during my three year tour in Germany and traded it in 1974 when it had 50,000 miles on it. At the time that was considered a lot of mileage.

[24] I followed up on this during my Germany tour and earned an associate's degree from the University of Maryland. Later I earned a bachelor's degree from New York State University in history.

[25] Criminal Investigation Command.

months. I guess Roy[26] never could find me an opening [with his police department]. At least I never heard from him.

We bought us some cane poles and went fishing the other day. Carol caught a 1 pound carp and was tickled pink. I didn't catch anything. They've got some real nice fishing places here. I'm going to do some more this weekend. As long as we just fish with worms we don't need a license.

We're all okay and I hope you all are too. Well, I've run out of things to write, so I guess I'll go for now. You all be careful and write soon.

                            Love,
                            Walt & Carol

P.S. Write soon.

Our quarters at Fort Campbell, Kentucky.

---

[26] Roy Myers, my first cousin on mom's side and a sergeant with the Cushing, Oklahoma police department.

SP/5 & Mrs. Walter L. Cross
4147A Lee Village
Fort Campbell, KY 42223

Mr. & Mrs. Earl L. Cross
Rt #1
Ripley, Oklahoma 74074

19 April 1970

Dear Mom, Dad, and Kids,

The first thing I want to say is thank Theresa and Kimberly for the nice birthday card. It sure is pretty and I really appreciate it. Well, today is Sunday and tomorrow I will legally be an adult.[27] Carol is in the kitchen making my birthday cake. She bought me a present and has kept it wrapped and in plain sight for the last week. I think she was just tempting me.

A sergeant that lives next door is getting a divorce so he is selling his furniture. I think we're going to get his table & chairs. They're Spanish [style] and should go with the rest of our stuff. Also we're thinking about getting his stereo. He wants $100.00 for it. And downtown they run $154.00 and he's throwing in all his records. Also with the dinette set he's giving us a rug.

There is something I'd like to say to you all. I know you all have never had a vacation. Now, with your camper you can.[28] So this is an invitation to pack your gear and come on out. We're just a day's drive away. And just a few miles from here is Lake Barkley and Lake Kentucky. We went fishing the other day and caught 19 fish, some cat and bass and a few sun perch. Carol caught the

---

[27] I turned 21 years old on April 20, 1970.
[28] Dad had recently bought a used camper for his pickup.

most of them. It's really pretty there and there are all kinds of places to fish.

We've got plenty of room. You and mom could sleep on our new bed and the girls could sleep on the two couches. Robin can sleep on the floor or we can fix something else up. I know you all would enjoy it and we'd love to have you all come. If you can get a few days off.

I haven't heard from Don and Di in quite a while. I guess they're mad at us for some reason. Well, not too much has been going on lately. Just the same old stuff. Here's hoping everyone is okay and write soon.

> Love,
> Carol & (legal) Walt

The black box on this and other awards that follow blocks sensitive information.

When I reported to post headquarters, the personnel sergeant, a Staff Sergeant Ryan Cornelius, sat me down at his desk and asked if I wanted to remain a medic or would I like to work for hospital personnel.

Not keen to work in a hospital after two years as a field medic I took him up on his offer. I spent the next six months working for Staff Sergeant Larry W. Cathy and received a maximum efficiency report from him. He wrote:

"SP5 Cross is one of the finest soldiers that I have had the pleasure of working with, and without exception is qualified for promotion to the next higher rank."

A month after receiving my first Bronze Star for Valor, I received a second, this one for meritorious achievement while serving in ground combat operations. Captain Madden, my battery commander, had recommended I be awarded the Silver Star Medal for Gallantry in Action. When his recommendation was downgraded to a Bronze Star for Valor he felt compelled to recommend the second Bronze Star for meritorious service. I would likely have received this award even if I had received the originally recommended Silver Star.

> CITATION
>
> BY DIRECTION OF THE PRESIDENT
> THE BRONZE STAR MEDAL
> FIRST OAK LEAF CLUSTER
> IS PRESENTED TO
>
> SPECIALIST FIVE WALTER L. CROSS, ███████
>
> HEADQUARTERS HEADQUARTERS AND SERVICE BATTERY, 1ST BATTALION, 7TH ARTILLERY
>
> 1ST INFANTRY DIVISION
>
> who distinguished himself by outstandingly meritorious service in connection with military operations against a hostile force in the Republic of Vietnam. During the period December 1968 to December 1969
>
> he consistently manifested exemplary professionalism and initiative in obtaining outstanding results. His rapid assessment and solution of numerous problems inherent in a counterinsurgency environment greatly enhanced the allied effectiveness against a determined and aggressive enemy. Despite many adversities, he invariably performed his duties in a resolute and efficient manner. Energetically applying his sound judgment and extensive knowledge, he has contributed materially to the successful accomplishment of the United States mission in the Republic of Vietnam. His loyalty, diligence and devotion to duty were in keeping with the highest traditions of the military service and reflect great credit upon himself and the United States Army.

Above is the citation for award of the Bronze Star Medal for meritorious service during ground operations against a hostile military force. The end date should actually read January 1970.

```
                    DEPARTMENT OF THE ARMY
                  Headquarters, 1st Infantry Division
                     APO San Francisco  96345

GENERAL ORDERS                                          15 October 1969
NUMBER   12917

                    AWARD OF THE BRONZE STAR MEDAL

1. TC 320. The following AWARD is announced.

CROSS, WALTER L ▮▮▮▮▮▮▮  SPECIALIST FOUR United States Army
Battery A 1st Battalion 7th Artillery

Awarded:          Bronze Star Medal with "V" device
Date of action:   9 September 1969
Theater:          Republic of Vietnam
Reason:           For heroism not involving participation in aerial flight,
                  in connection with military operations against a hostile
                  force in the Republic of Vietnam: On this date, Specialist
                  Cross was serving as a medical aidman with his unit when the
                  friendly encampment was suddenly subjected to an intense
                  rocket-propelled grenade barrage followed by a massive human
                  assualt which caused numerous casualties. With complete
                  disregard for his personal safety, Specialist Cross left
                  his relatively secure position and maneuvered through the
                  hostile fusillade to the aid of his fallen comrades. After
                  administering first aid to the injured personnel, Specialist
                  Cross organized the prompt removal of the casualties to the
                  medical evacuation zone. His courageous initiative and
                  sefless concern for the welfare of his fellow soldiers were
                  instrumental in saving several friendly lives. Specialist
                  Cross' outstanding display of aggressiveness, devotion to
                  duty, and personal bravery is in keeping with the finest
                  traditions of the military service and reflects great credit
                  upon himself, the 1st Infantry Division, and the United
                  States Army.
Authority:        By direction of the President, under the provisions of
                  Executive Order 11046, 24 August 1962.

FOR THE COMMANDER:


OFFICIAL:                      A. G. HUME
                               Colonel, GS
                               Chief of Staff

J. P. BOTT
First Lieutenant, AGC
Assistant Adjutant General
```

The orders and citation for the Bronze Star Medal for heroism in combat.

Orders were soon produced assigning me an MOS[29] of 71H20, personnel specialist. My medical specialty was shoved downward to become my secondary MOS.

Carol and I enjoyed the life at Fort Campbell and made the best of the nearby recreational area called 'Land Between the Lakes' where I taught her to fish, a hobby that has lasted her a lifetime. We also became bold and bought our first new car, a 1970 competition orange, Plymouth Duster. I loved that car and later took it to Europe and drove it on the famous autobahns of Germany.

Meantime, I received my honorable discharge at Fort Campbell in October of 1970. During that ceremony I was presented a citation from the post's commanding general, Brigadier General William H. Birdsong that reads:

"Specialist Five Walter Lee Cross is cited for exceptionally meritorious service from February of 1970 through October 1970. During this time he maintained a succession of positions with the Personnel Division, United States Army Hospital, and Fort Campbell, Kentucky. From the time he

---

[29] Military Occupational Specialty

arrived Specialist Five Cross utilized every opportunity to broaden his experience and job knowledge. Within a very short period of time he demonstrated exceptional proficiency. Not being satisfied to learn only the fundamentals, he learned the underlying reasons and functions pertaining to his specialty. On many occasions he worked beyond normal duty hours to gain experience and to accomplish the mission in a usually understaffed situation. The military and civilians who worked in the Personnel Division depended heavily on his ability. Later, Specialist Five Cross took on additional duties and fulfilled them with unparalleled competence. His wide experience, careful attention to detail and ready acceptance of responsibility contributed immeasurably to the Personnel Division. His outstanding achievements are in keeping with the highest traditions of the military service and reflect great credit upon himself and the United States Army."

> *William H. Birdsong*
> *Brigadier General*
> *Commanding*

General William Birdsong Fort Campbell 1970.

Chapter VII
Overseas to Europe

I toyed with the idea of leaving the service and did for a whole eleven days, discovering in that short time that I preferred the military to civilian life. I reenlisted with a guaranteed assignment to Germany for three years, and a six-thousand dollar re-enlistment bonus, which was a very great deal of money to a twenty-one year old in 1970.

I applied for concurrent travel of dependent so Carol could come to Europe the same time I did, and we found ourselves temporarily assigned to Fort Polk, Louisiana while the request was processed. I had come full circle, and was back at the post where I had taken basic training. We were there for about three months living in the nearby town of Deridder. During that time I collected half of my reenlistment bonus from Post Finance. I carried it in a briefcase to the bank for deposit; I'd never had so much money all at once.

When we got to Louisiana Carol told me something odd. The FBI called her mom's house and asked to speak to me. When they learned I had gone back in the Army the special agent calling said "Too bad, he wanted a job with us. Please tell him if he gets out of the Army to call us."

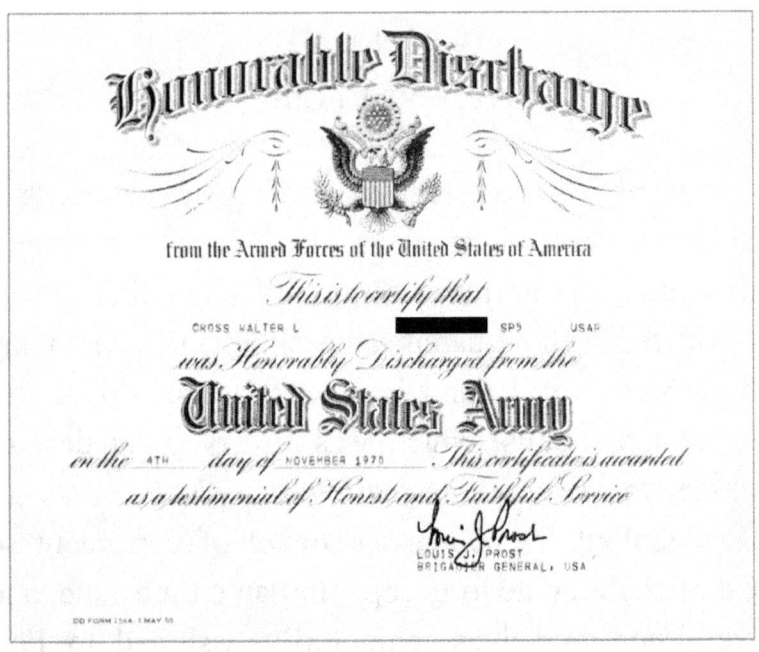

Honorable Discharge in November of 1970. The designation of USAR (U.S. Army Reserve) is because after my three year enlistment on active duty, I still had an additional three year obligation to the military.

This is a mystery to me because I never applied to the FBI.

Concurrent travel was denied, so in January of 1971 I flew aboard a USAF troop transport plane to Germany. Instead of normal seats we sat in web seats facing backward. Carol was to follow within 90 days. I had a housing assignment code of ALPHA THREE GREEN which meant that upon Carol's arrival in March, we would be provided government housing. And we were assigned quarters in the John F. Kennedy housing area.

My orders were to report to Company A, 237th Combat Engineer Battalion stationed at Wharton Barracks in the large city of Heilbronn, Germany about twenty miles from the even bigger city of Stuttgart. I was still a personnel specialist, but my duty position title was 'company clerk'. It was a position I was not destined to remain in for long.

When I arrived in country and reported to the commanding officer of Company A, Captain David Shockey, the 237th Engineer Battalion (Combat) had been in Europe and Germany for seventeen years, since D-Day on the 6th of June 1944.[30] Originally activated at Fort Carson, Colorado as the 2nd Battalion, 49th Engineers in August of 1942, it soon became an independent battalion bearing its new designation as the 237th.

Led by Lieutenant Colonel Herschel E. Linn, the battalion was attached to the 1106th Combat Engineer Group, VII U.S. Army Corps. It would serve in its combat support role across France, Belgium, and Germany in five campaigns designated *Normandy, Northern France, Rhineland, Ardennes Alsace, and Central Europe.*

---

[30] After its first year in Europe the battalion was sent home and then returned to Germany in 1955.

Landing on Utah Beach, the 237th with naval elements and Company B, 299th Combat Engineers attached, joined the 70th Tank Battalion in the assault on Hitler's "Fortress Europe". This provisional battalion operated in support of the 82nd and 101st airborne divisions. The plan was to land following the infantry but in fact the battalion landed with the infantry.

Despite the confusion of battle and intense enemy fire, the men of the battalion waded ashore carrying back packs loaded with sixty pounds of explosives as well as their personal equipment and weapons through water as deep as four feet. They then encountered an enemy mine field which they cleared to the depth of a thousand yards before the tide began to rise. For this action the unit received a citation from the VII Corps which later became a Presidential Unit Citation[31] from then President Franklin Delano Roosevelt.

During the action the battalion drove the Germans inland and materially contributed to the success of the invasion of Europe. In recognition of this the French Government awarded the 237th Engineer Battalion the French Croix de Guerre with Silver Star.[32]

---

[31] General Orders 88, War Department, 23 November 1944 Tab D.
[32] General Orders 43, War Department 19 December 1950 Tab D.

The battalion was known for its speed and courage while constructing bridges, railways, and canal bypasses as well as clearing enemy minefields, and marking and repairing roads. The battalion would also bridge the Rhine River and on more than one occasion, be used in its secondary mission of infantry. Men of the battalion would receive many valor awards including one Distinguished Service Cross, sixteen Silver Stars for gallantry, fifty-eight Bronze Star Medals, a Soldiers Medal (for life saving not in combat), and three personal awards of the French Croix de Guerre. During combat thirty-five soldiers of the battalion were killed and a hundred and twenty-eight wounded, as well as 3 missing in action. presumed killed. This was the proud unit I joined in January of 1971.

Atomic Cannons at Wharton Barracks

A number of years before my arrival there was an unusual and interesting unit assigned to Wharton Barracks besides the 237th Engineers.

That unit was Battery B and C of the 867th Field Artillery (Atomic Cannon) Battalion. In the words of Lieutenant Frank Hubp who served with B Battery:

"These big heavy guns could shoot an atomic round (shell) 20-25 miles – and were mounted on a huge carriage unit that was pushed and pulled at the same time by two trucks [one] at each end. While out on maneuvers, the two drivers would have to communicate with each other constantly to get through the narrow streets and traffic.

An M65 280mm cannon, known in the vernacular of the time as "Atomic Annie" sets on the Battalion Quadrangle in front of the unit headquarters building. A note on the back of this 1956 photo reads "McFall is standing on the gun's rail".

But sometimes we would cut it too close and take a little corner off buildings. We actually had a special unit that went ahead of us to test whether the bridges could handle our weight. The Russians

BOOMER SOLDIER

Two pictures of the cannon's hauler.

weren't allowed to be in our area, but every once in a while we would spot a sedan of the Russian military – and we'd have to call the MPs (military police). This was the time when you never knew if and when the Russians were going to attack."

The 867th Field Artillery Battalion was one of four such battalions assigned to the 42nd Field Artillery Group. Other assigned battalions were the 59th FA, 868th FA, and the 264th FA. The battalions, consisting of three batteries, were broken down into individual two-gun batteries and assigned throughout Army bases in such towns as Baumholder and Heilbronn. The 42nd also had Corporal Missile battalions that were comprised of nuclear armed tactical ground to ground missile batteries. These, combined with the atomic cannons comprised the Army's nuclear deterrent for many years.

Above is the unit crest (insignia) of the 867th Field Artillery Battalion. The unit motto at the bottom of the crest is *EXSTINCTOR* meaning to make extinct. I think it's the appropriate motto for an atomic weapons unit. I have one of these rare crests in my personal collection. The M65 remained a deployed U.S. Army atomic weapon until 1963.

The actual nuclear explosion of an M65 shell.

On May 25, 1953 at 8:30am, the atomic cannon was tested at the Nevada Test Site. It resulted in the successful detonation of a 15 kiloton shell at a range of seven miles. This was the first and only nuclear shell to be fired from a cannon.

After the successful test, there were at least 20 of the cannons manufactured at a cost of $800,000 each. They were deployed overseas to Europe and Korea, often continuously shifted around to avoid being detected and targeted by opposing forces. It remained a prestige weapon serviced by the best trained artillerymen and was not retired until 1963.

Chapter VIII
Company A, 237th Engineer Battalion

In the spring of 1971, not long after my arrival in Company A, the unit hired a photographer for pictures. You can see my specialist five rank, a rank designation no longer used in the Army, the 7th Engineer Brigade shoulder patch, and the 237th Engineer Battalion (Combat) unit crests on my shoulders. Photograph taken in Heilbronn, Germany.

One of the primary duties of the company clerk is the preparation of the unit's 'morning report' a daily accounting of the unit's activities and the status of its assigned personnel. It absolutely, positively, has to be correct. The morning report goes up the chain of command to the 7th Engineer Brigade commander, the 7th Corps commander, and on to the commander of United States Army European Command and eventually to the Pentagon and later into the national archives.

The company commander's career depends on accuracy of the morning report, and the company clerk prepares the morning report under the direct supervision of the company first sergeant. There is a lot of pressure associated with this task. But after the report is done, the first sergeant satisfied, and the company commander has signed off, the rest of the day is pretty easy. I excelled at the job, for which I would be 'rewarded' by reassignment out of the company.

On the Heilbronn base, known as a kaserne; German for barracks, was a special engineer unit, special and secret, and out to recruit me.

Not long after settling in to Company A the first sergeant informed me I had to report to the ADM Platoon for a mandatory interview. He went on to explain that anyone assigned to the battalion who had a "general technical" (GT) score of 110 or higher were required to be interviewed. My GT, which correlates roughly with an IQ score, is 125. Not trained as an engineer I asked him what an ADM platoon was.

He told me that ADM was an acronym for 'Atomic Demolition Munitions'. He went on to say that the ADM Platoon actually trained with and handled small tactical atomic weapons for use in destroying bridges, mountain passes, dams, autobahn interchanges, and other strategic choke points upon a Warsaw Pact invasion of Western Europe.

The ADM Platoon patch.

The platoon was actually assigned to VII Corps but was attached to the 237th for administration and support.

I went to the interview and the platoon sergeant asked me if I wanted to be special and exempt from all duties such as guard detail or work details (which I was already excused from as the company clerk) and draw hazardous duty pay. Just to mess with him I asked him to explain what I would be doing and he told me he could not explain until I agreed to become a member of the platoon. Of course, I declined. But it was only the first interview I would be subjected to.

Left is the 7th Engineer Brigade shoulder patch. The shield is red and white with a black saltair cross superimposed and a gold castle turret over all. Right is the 237th Combat Engineer Battalion unit crest. The motto reads "Dedicated and Determined". It depicts a white snow covered mountain on an engineer red background.

In the spring of 1971 a terrible event took place, an event that would shake the command and reach high into the Army's establishment itself.

The German girlfriend of one of our engineer soldiers came by the barracks to speak to him. This was something she had done before with no problem. But this time it was dark. There were two entrances to our barracks building, one door led to our Orderly Room and at that time of day, early evening, it was manned by a company NCO who was on CQ or 'charge of quarters' duty. It was his job to supervise the building overnight and report any problems to the battalion officer of the day. He was assigned a junior enlisted man as an assistant known as a CQ runner.

The other entrance led directly into the sleeping quarters of a completely different unit. This unit was not a part of our battalion and their portion of the barracks was sealed off physically, from us. The German girl approached and asked to speak to her boyfriend. A soldier standing just inside the door told her to wait and he went into the barracks and told two other men that a German prostitute was at the barracks door looking for some action.

All three approached her and the girl tried to back away and out of the entryway. The three grabbed her and dragged her, protesting, into the barracks. Soon, there began a gang rape of this poor girl that eventually involved nearly every man in that barracks! It was not until morning that it was discovered what had happened. The girl was in total shock, injured and abused in a manner almost beyond belief to include rape, sodomy, and oral sodomy.

The local CID (Criminal Investigation Division) of the Heilbronn Agency got involved and during the discussion on how to conduct a line up to identify her assailants, I suggested the Company Orderly Room be used. With the blinds drawn and the sun shining on them from the outside, the girl could see out, but the suspected men couldn't see in. Some days after the assault, the lineup was conducted, and although she could not hope to identify all of them, she did manage to designate many of her principal abductors including the soldier who tricked her into waiting until he returned with his accomplices.

There followed many a courts-martial, dishonorable and bad conduct discharges, and imprisonment at Fort Leavenworth. This included one man who was reportedly the son of an undersecretary of either the defense department or the Army. It was during this affair that I came to the attention of the Heilbronn Agency of the CID.

BOOMER SOLDIER

Above is a picture of me and some of the men in my headquarters section taken during an FTX (field training exercise) in a forest near Volkach, Germany on September 17, 1971. Behind us is our tent. We spent thirty days in this very cold forest. The soldier on the far right is Patrick McArthur from El Centro, California, he was an amateur photographer, and he both developed and printed this photograph. We worked at the time for an adjutant, the S-1 officer (a captain) who was a real asshole. He had been caught up in a RIF (Reduction in Force) and was to be discharged for bad officer efficiency evaluations. He took his frustrations out on our section. I tried to mitigate his negative effect on the men as best I could as section chief. The other men from the left are John Puckett of Clinton, Kentucky and Thomas McDaniel from Colorado (standing to my left). The captain has his hand resting on Puckett's helmet; McArthur thoughtfully cut him out of the photograph.

The coldest day I remember in Germany, riding in the back of an open jeep from the Grafenwöhr training area near the East German border. This was in the winter of 1972; exactly 28 years after my dad experienced almost the same conditions near this same area during World War II. Dad was an armored cavalryman and fought in the Battle of the Bulge. I wonder what that document is in the pocket of my field jacket. I would guess a map as I was the senior soldier in the jeep.

Carol and I enjoyed the assignment very much, despite the field problems I had to attend that sometimes lasted weeks at a time. These exercises usually consisted of camping out in a dark, cold German forest while we conducted a simulated combat operation, it was good training.

In our free time Carol and I traveled around Europe, visiting Paris, London, and wide open Amsterdam. We even spent two weeks on the shores of the Mediterranean Sea in Spain in a little town named Tossa de Mar (gold of the sea), and watched the bullfights in Barcelona. Carol, always lucky in contests, had won a two week, all expense paid trip to Spain.

Upon my return to the unit I was called to battalion headquarters for an interview with the battalion commander and his adjutant. I wasn't sure what it was about and was a bit nervous when I reported.

A favorite place of both Carol and I in Germany is the ruins of Winesberg Castle, one of the oldest castles in Germany and located just a short distance from Heilbronn. Carol took this photograph of the castle turret on a cold winter's day in 1972. You can see the frost on the trees. The view from the castle is sweeping and steam driven trains pass below in picturesque display. Outfitted with wind guitars in its windows, the castle often moans hauntingly with the evening breeze. Carol won a 1st prize ribbon for this photograph in a local county fair.

The battalion legal clerk, that individual charged with preparing both non-judicial and court-martial paperwork, had rotated back to CONUS (continental U.S.) and there was no replacement in sight. This time there was no decision for me to make, I was selected to be the new legal clerk and that was it. I changed my duty location across the quadrangle from Company A to the battalion headquarters and continued to march. My meticulous preparation of the "morning report" now transferred to the legal matters that also had to be absolutely correct.

Here is another photograph by Carol also taken during the winter of 1972. This is the medieval town of Rothenberg Ob Der Tauber (on the Tauber River). She really is a good photographer and this photograph, as well as the earlier one, still graces our home to this day. It is even more impressive in color.

During my tenure as the company clerk I was rated twice on my efficiency, the first was in June of 1971 after I had been assigned for six months. First Sergeant Charlie E. Thomas wrote my report and stated:

"SP5 Cross is one of the better young soldiers it has been my privilege to work with. His enthusiasm and dedication to duty are to be commended. SP5 Cross will develop into an outstanding non-commissioned officer and be a credit to himself and the United States Army."

I really liked First Sergeant Thomas and found him amusing at times. But one of the things I found amusing, the command did not. During our morning formations at o'dark thirty in the mornings, the first sergeant would take the opportunity to berate the company. His favorite phrase was "…you dirty sister raping bastards!"

His fame for colorful language was well known and crowds, usually consisting of dependents, would gather on the hill above our company assembly area to listen. I recall one morning when he was really laying in on the company, he paused, looked up and pointed at the dependents on the hill, turned to Captain Shockey and said "And another thing sir, those goddamned dependents should be banned from anywhere near the fucking company area!"

With a little urging from the command he decided to retire not long after and Carol and I had him over for a farewell dinner at our quarters in the John F. Kennedy housing area. After the meal and a couple of exceptional German beers, he presented me with his first sergeant pins and told me I would need them in the future.

The sleepy little fishing village of Tossa de Mar, Spain on the shores of the Mediterranean Sea. Carol and I enjoyed an excellent two week vacation here, one that even rivaled our trip to Hawaii three years before. Note the Roman era tower in the background as it is the prominent feature of the next photograph.

Years later when I was promoted to master sergeant I wore those pins (after first carefully removing the diamond of a first sergeant from the middle of the stripes, master sergeant and first sergeant both have three stripes above three rockers and are the same pay grade of E8).

I remember with great fondness that when I first arrived First Sergeant Thomas took me to the nearby small village of Stetten where I had my first real German cooked meal. It came with kalbsshnitzel (breaded beefsteak) pomfritz (French fries) spaetzle (dumplings) kartoffelsalat (potato salad) with gravy, and a large bier (beer) in a ceramic and wire flip-lid re-sealable glass bottle. It was delicious and a great deal of food! I could only eat about half of the meal.

However, by the time Carol arrived a couple of months later and I took her to this same gasthaus, I could eat all of my food and half of hers! She was delighted by the small restaurant despite the fact it was situated above the barn where the family kept its farm animals! It became our favorite place to eat and we made the short drive to Stetten often during our time in Germany.

My next efficiency rating was done by the acting first sergeant, Sergeant First Class John M. Somers only three months later when I transferred on the first of September to the battalion headquarters. Somers gave me the highest ranking of "Outstanding" in five areas and "Excellent" in one, I never received a copy of the finished rating and so don't have his comments.

Chapter IX
Friends; Human, and Animal

During our stay in Heilbronn Carol and I had become friends with another couple, Rick and Carol Palms.[33] Carol was from Chicago and Rick hailed from the Chicago suburb of Mundelein, Illinois.

Carol was a lovely woman both in appearance and actions with an air of gentility about her. She always comported herself like a lady and had a very droll and dry sense of humor that often snuck up on you. I later found she was a Butler before her marriage to Rick. The Butlers are an old and wealthy American family whose fortune was made in aviation. Her grandfather, Paul Butler, founded the Butler Aviation Company in 1946 just after WWII in Chicago. During its history, Butler Aviation supported the flight needs of many organizations and individuals, including US presidents, international government leaders, celebrities, major corporate executives, and airlines. It remained independent until 1987 when it was sold.

Rick in contrast came from a working class family from Michigan. Carol and I visited his parent's home after returning from Europe. A very pleasant couple, they introduced us to the agreeable

---

[33] Not their actual last name.

custom of drinking Japanese plum wine with hot tea or coffee. Rick's dad was a WWII Army ranger, and lost several toes to frostbite during the Battle of the Bulge. He confessed to me that the members of his ranger company would not accept the surrender of SS troopers and took none of them as prisoners.

Mouse, the Cat

Rick and Carol owned a couple of cats. One was a white Persian tomcat named Leonardo de Fluff. The other was an American shorthair, sometimes called an alley cat named Scat. Scat was as unfriendly as Leonardo was friendly. We owned our own cat affectionately known as Mouse. Mouse was young and very inquisitive as cats are at that age. We took him with us to visit the Palm family cats one day.

Mouse seemed to get along with Leonardo pretty well, but Scat would have nothing to do with him. We sat in their small living room talking and enjoying refreshments while Rick set up his small HO gauge railroad mounted on thin plywood and sitting on the coffee table.

He slowly ran the tiny electric train around the track that meandered through a miniature German village. Mouse was fascinated and after watching the tiny train a couple of turns around the circular track he hopped upon the coffee table and straddled the track.

When Rick saw him in this somewhat vulnerable position he got the bright idea to turn the train up on high speed. As the engine, trailed by a number of train cars lurched ahead Mouse crouched down in alarm, his eyes glued to the moving snake-like creature speeding toward him. As the trained neared he suddenly jumped straight up in the air!

I started to laugh as the impetus of his leap shoved the entire plywood train and village setup off the coffee table and sent it crashing to the floor and knocking Rick's glass of wine on the living room rug.

For a moment Rick was angry, but I reminded him it was his entire fault. Eventually he joined with me and my Carol and his Carol in laughing it all away.

Mouse was an American shorthair cat. Gray and white in color and with a particular mischievous bent to his personality. He also had a flaw, his voice box wasn't quite right and he couldn't meow like other cats. When he tried the sound that came out was like a stutter that sounded something similar to "ka…ka…ka…". Although a little disconcerting, that was the best he could do. Over time Carol and I grew used to it and wouldn't have traded his sound for even the sweetest of 'meows'.

The infamous no-meow cat, Mouse, when just a kitten. He was the only cat I ever knew that tolerated a leash. He would grow up to be an excellent mouser.

I was already writing in my spare time, or perhaps I should say attempting to write. I had a portable Olympia typewriter with a neat case that opened up and allowed me to type without taking the machine out. Of course the clatter of the typing keys interested Mouse and he came to investigate. He hopped up on my desk and watched a moment, then boldly stepped into the space between the case and the typewriter itself. I knew what was on his mind as he watched the keys fly.

For the first few moments Mouse sat contentedly, intensely watching the keys rise up and slap the paper with a sharp "clack, clack, clack" sound. I knew what he was going to do and I warned him "Mouse, don't you do it!" But he did, he stuck that inquisitive paw of his inside the mechanism to catch that clacking key! Just as he did of course two keys

(I am a pretty fast typist, thanks to taking a typing class in high school) smacked him resoundingly on the paw before he pulled it back with a squall and a leap straight up in the air! I laughed so hard I bent over double and Carol came running in to see what happened and laughed too when I explained.

Meantime, Mouse had retreated to his favorite hiding place behind the couch and didn't come out for a good half hour. When he did he took up a position beside my rubber tree plant. Carol had cultivated it for me and in his cat brain Mouse knew it was mine. He sat on his haunches and looked at me sullenly. I looked back, still laughing at him and he did something that still surprises me to this day. He stood up, glanced at me over his shoulder and slashed the nearest fat leaf of my rubber tree! There is no doubt, Mouse had gotten his revenge!

I jumped up and exclaimed "You rascal!" Mouse took off like a shot, got around me and quickly disappeared under the couch again. I chuckled and shook my head in amazement. Soon I was back to writing and eventually Mouse come back out from under the couch. He eyeballed me for a time and then sauntered back over, jumped up on my desk as if nothing had happened and took his seat in the case lid.

Unfortunately I disposed of my Olympia (made in West Germany) many years ago. This picture shows Mouse in his favorite sitting area as I am typing. The lid lay back, leaving a very inviting place for Mouse to sit in. Mouse was now a grown up tom cat and is sporting his U.S. Army 'dog' tag.

This became a routine for him, but he never stuck his paw in the machine again, although he loved to watch the keys smash on the paper and never tired of the carriage being returned. When I finished for the day I would stack my papers and that was his clue to get out of the lid. He was my constant companion during all of my writing.

We got Mouse as a kitten just before the onset of the German winter and so for the first several months of his life he was an apartment dweller. However, that did not quell his instinct to hunt and there were only two things in the apartment he could hunt, the occasional fly and, Carol.

He was death on flies and would stalk a single insect throughout his domain until he assassinated it through sheer perseverance. But each night at bedtime, he became a big cat of the jungle. Just before Carol and I would head to the bedroom he always took up a strategic position to ambush his prey.

I, being the alpha male of the domicile would saunter across the bedroom and get into bed. But Carol would peer into the room looking for Mouse although she rarely saw him until she made a dash for the safety of the bed. That is when the king of beasts, at least in our apartment, would make his appearance. He would grab her by the ankle and hold on while he nipped playfully, but sometimes painfully at her shin and ankle. Carol would yowl and endure it until she reached the bed where she could shake him loose.

This was a nightly occurrence. He had only tried it on me one time while he was perfecting his technique. It was the wrong move! Startled, I shook my leg pretty hard when he latched on and he careened into the bedroom wall! It was unintentional on my part but it did teach Mouse, that like the water buffalo for African lions, I was not easy prey. He never tried to take me on again.

He did something one day that could have had repercussions for my military career had my captain taken umbrage. Mouse, always looking for something new to occupy his time, came up with a new exercise. Somehow he discovered that if he ran fast enough down the hallway that led to our apartment front door, he could bounce off the walls, gaining height with each leap until he was chest high to a human. Carol and I watched him do this a number of times, and since it didn't cause any damage and he enjoyed it, we didn't try to make him stop. That turned out to be a mistake!

To make a little spending money Carol would occasionally babysit for the children of soldiers in the battalion. One day she kept Captain Shockey's young daughter. When the captain's wife came to pick up her daughter she stood talking to Carol in the doorway.

Down the hall came Mouse, running and bounding from wall to wall. Carol and Mrs. Shockey stood in shock, no pun intended, and watched him come. He ended what must have been a somewhat terrifying event for the captain's wife by landing on her chest, claws bared, and slowly slid down the front of her knit jacket, snagging long rows in the fabric as he went. Carol quickly plucked him off, apologizing profusely, but the knit suit was, of course, ruined. The terror cat had struck! Luckily, the captain never mentioned the incident to me although Carol offered to pay for the jacket.

That spring, following six months inside the apartment, we took Mouse outside for the first time. It was a warm, beautiful and sunny day, a real relief for the many months of continuously overcast skies. His initial reaction to once again experiencing the out-of-doors was a little surprising. He couldn't get over the wide open spaces and the vaulting sky. He was so used to being surrounded by walls and a ceiling. He hunkered down, literally crawling about in the green grass on his belly and looking up myopically at the scary blueness above with its white puffy objects that continually changed shape.

A passing group of German children stopped and pointed at Mouse's strange antics and asked:

"Vas ist der katze namen?" I replied "Das namen der katze ist Maus."

They turned to one another in amazement and exclaimed "Das namen der katze ist Maus!" And they all laughed merrily, continuing on their way to school, glancing back and giggling as they talked and joked among themselves.

My wife Carol (left) and Carol Palms sitting outside the gate to Rheinstein Castle in Germany.

Rick was a chaplain's assistant, and his function was to assist the chaplain in church services whether at the post chapel on Wharton Barracks or in the training field. He drove the chaplain's jeep and was his armed guardian during combat as the chaplain; a man of the cloth of course, carried no weapon.

Walt, Rick and Carol Palms at the gate to Rheinstein Castle.

Rick was a very good friend; he and Carol gave me gifts at Christmas that I have kept and treasure to this very day, to include an exceptional chess set and a German beer stein personalized to show my service with the 237th Engineer Battalion. He and I often played war games and sometimes Rick, Jerry Shaw, and I got together to play three-handed war games. But, Rick had a dark side to him that I did not learn about until our return stateside. It seems that while he was the post chaplain's right hand man,

Rick and Carol Palms in Dinklesbuhl, Germany.

he was also having an affair with the chaplain's wife!

Soon after his discharge he divorced Carol and married the chaplain's wife who had also divorced. We were pretty shocked and saddened for Carol. But like the lady she always was, she soldiered on.

For a number of years we kept in touch with Carol but not Rick. She later married a college professor and became Carol Bushong. She completed her own education and got a doctorate while having a family of three children, a boy and two girls. The girls are as beautiful as their mother.

A beer stein presented to me by Carol and Rick Palms. Depicted in full color is the 237th unit crest and flag personalized to me. The reverse has the arms of Heilbronn, Germany with the years of my service 1971 through 1973. I treasure this gift as a reminder of friendship and of my military service. It shows my rank as sergeant because by this time I was a sergeant.

Here is another Heilbronn Beer stein depicting the town clock. I actually didn't buy this in Germany I found it offered in an estate sale in Stillwater, Oklahoma. What are the odds? Rather than hand painted like the one Carol and Rick got for me, this picture was obviously a print glued to the stein and lacquered in place. But it is a reminder of a happy time for Carol and I in an enchanted and memorable country.

This is my very good friend Gerald Shaw, who went by the nickname of Jerry. He and his wife Sue were constant companions of Carol and I. Here we are playing a war game in my quarters on Christmas day in 1972. Jerry took my place as the battalion legal clerk, in a move that proved very good for his military career and his retirement several years later.

Carol and Sue review the defeat of Napoleon at Waterloo, and a bunker outside the Belgian town of St. Vith. It was at St. Vith my father, PFC Earl L. Cross, and his unit, the 87th Cavalry, 7th Armored Division, experienced their toughest fighting during the Battle of the Bulge.

It was down this road, leading from Germany into Belgium the Germans attacked my dad and his unit in St. Vith. There was a lot of hard fighting for both sides on this road.

St. Vith as we saw it on a cold and wet day in 1972, the scars of battle long gone. The town's signs are in French and German to this day.

Carol took this picture of Sue Shaw and I inspecting a monument to the men who fought at St. Vith. As can be seen by the large snowflakes in the frame the weather suddenly changed. And this was in the month of May, 1972.

Our good friends Sue and Jerry Shaw out to dinner with us. Sometimes the toll taken on servicemen and their families is just too much. Unfortunately Sue and Jerry would later divorce.

After more than five months of working for the colonel who signed court-martial memos with a purple pen as the "Purple Avenger", it was time for a change. I had seen the seamy side of the battalion setting in as the recorder during court-martials, and writing charges for non-judicial punishment. The unit was afflicted with racial tension that often led to fights, drug abuse, and other problems. On one occasion a number of soldiers damaged the radiators of their company trucks to avoid having to go on an FTX[34]. Part of their punishment was to march several miles to the training site instead of riding in

---

[34] Field Training Exercise.

vehicles. Captain Richard Ritchie, the battalion adjutant wrote my efficiency rating and stated:

"SP5 Cross is intelligent, hardworking, and conscientious. He has been utilized in a number of positions at the battalion headquarters and has performed well in all of them…"

However, his scoring was only two areas "Outstanding" and the rest "Excellent". This brought a sharp rebuke from the battalion executive officer Major Donwell D. Whitley Jr. who wrote:

"I do not concur with the rating officer for the following reasons. SP5 Cross has a better than excellent attitude toward the U.S. Army and any task assigned to him. Additionally, although not in a leadership position, his natural leadership among his contemporaries and his sense of responsibility toward his job has been outstanding."

I had never seen before, nor since, where an indorsing officer had over ridden a rating officer, at least not on any other efficiency report I ever received. The major gave me all outstanding ratings.

Before the battalion would let me leave, they asked me to recruit my own replacement as the battalion legal clerk. A good friend of mine, a combat engineer, Sergeant Jerry Shaw[35], expressed an interest in learning how to be a legal clerk. He was a natural. I spent several weeks training him, while I worked with Staff Sergeant Phil Patton, the battalion career counselor, learning to become a reenlistment NCO.

After 90 days of OJT, like at Fort Campbell, I was given a new MOS, this time OOE career counselor/recruiter and laterally appointed from specialist five to sergeant. Staff Sergeant Patton gave me an efficiency report in March of 1973 that read:

"Sergeant Cross performs his duties diligently, in my opinion he is the best in his grade and MOS…I feel he should be promoted to the grade of staff sergeant E-6 immediately."

Ah, I hear you saying, that's how you became a recruiter. Yes and no.

As I was becoming a reenlistment NCO, the last combat soldier in Vietnam, Sergeant Max Beilke was departing Saigon in March of 1973. Ironically, years later, the now retired Master Sergeant Max Beilke was killed on September 11, 2001 (911)

---

[35] Jerry would go the opposite way from me with a lateral appointment from sergeant to specialist 5 and eventually to the highest rank of specialist, specialist 7, before becoming a legal field warrant officer. He eventually retired with the rank of chief warrant officer.

during the airplane terrorist attack on the Pentagon in Washington D.C.

Back in the United States, the last man to be drafted reported for training in June, 1973. The draft had ended; the last combat soldier in Vietnam had come home. However it was not until 2011 that the last man drafted who decided to make the Army his career, retired from active duty. Change was coming to the U.S. Army and men were needed to make that change take place as Army recruiters.

Sergeant Max Beilke is patted on the shoulder by South Vietnamese official Bui Tin as he prepares to depart at Ton San Nhut airbase. "I hope you return as a tourist…" he said "you are certainly welcome." But Beilke was killed in the attack on the Pentagon in September 2001.

Chapter X
Terrorism Rises in Germany

Advent of the Baader-Meinhof Gang

The left wing terrorist group Red Army Faction (RAF) started out as the "Baader-Meinhof Gang" during the year 1970 in Germany. It was one of the biggest series of events that took place while Carol and I were in Europe. It later took on the title Red Army Faction.

Despite killing 34 people, Baader-Meinhof garnered a degree of support from the West German population. Accordingly the militant group began to be accepted, if not outright admired by a number of Germans who disagreed with their nation's association with the American war in Vietnam.

Baader-Meinhof seized on this growing public sentiment and cultivated their image as outlaws authentically acting out their desire to break through fascism, just as Che Guevara was seen to break through American imperialism. Mimicking their American counterparts, the group began to use similar phrases in their speech such as "burn baby burn, off the pigs" and "right on".

German Landkris Polizi[36] manned street corners in Heilbronn and other cities carrying military style machinegun rifles. The gang made headlines and prime time newscasts in Europe and America.

After an intense manhunt, gang members Baader, Ensslin, Meinhof, Holger Meins, and Jan-Carl Raspe were eventually caught and arrested in June 1972. They were held in the newly constructed maximum security Stammheim prison located in north Stuttgart, only a few kilometers from our home in Heilbronn.

The Red Army Faction continued to be a thorn in the side of German authorities for more than twenty years after I rotated home from Germany. The

---

[36] State Police.

history of this militant group is interesting and both Carol and I vividly recall their advent.

The Red Army Faction no doubt influenced the creation of a similar group in Italy in 1970 known as the *Red Brigade*.

The other big and violent event took place only two months later in September of 1972. It would become known as the 'Munich Massacre'. And, it would include a demand for the release of Andreas Baader and Ulrike Meinhof of the Red Army Faction.

In the very early morning hours of September 5, 1972 eight dark clad members of the terrorist group *Black September* stole into the Olympic Village in Munich and kidnapped nine Israeli athletes. During a botched attempt at a rescue by German authorities all of the hostages were killed and five of the eight terrorists were killed. Three were taken to prison to await trial, but a high-jacked Lufthansa airplane won their release. The fate of the three is debated to this day, some say they died at the hands of the Israeli Mossad[37].

Soon after I received my new MOS as a retention NCO I was approached by yet another agency. During my time as a legal clerk I had found it necessary to speak to the agents of the Heilbronn CID Agency. As I mentioned they also knew me from the case of gang assault of the German girl.

There was a push at the time to get new CID agents from among serving soldiers. Through the base education center I had recently completed the requirements and received an associate's degree in science from the University of Maryland. It seems I was the kind of soldier they were looking for and I accepted their offer of becoming what they called an "agent in training" or *bootleg* agent. I spent the next year working in civilian clothes as a CID agent.

When I served in the CID, the command was undergoing a major reorganization. In September of 1971 Melvin Laird, the Secretary of Defense, directed the formation of the CID into a major Army command in and of itself. All investigative assets were brought under central control and placed under the command of Colonel Henry Tufts.

---

[37] Israeli Intelligence Agency.

In addition to this reorganization the CID was also struggling with the biggest war crime investigation of the Vietnam War. No, I'm not talking about the *My Lai Massacre*, but a war crime much larger in scope and one that few people know about because it was covered up, intentionally.

Tiger Force 1967

Ybarra 2nd row third man from left.

In February of 1971 Sergeant Gary Coy told Army investigators that he knew about a soldier in the 1st Battalion, 327th Infantry, 101st Airborne Division that murdered a Vietnamese baby in November of 1967. Specifically, he was accused of decapitating a baby. The case was designated *Number 221 the COY ALLEGATION.*

The allegation was already more than three years old and the Army didn't really know what to do with it. It was not seriously looked at until over a year later in March of 1972 when it came to the attention of Colonel Tufts after the CID's reorganization was well underway. That is when the CID began the search to find the men of *Tiger Force*.

It was soon discovered that the soldier accused of murdering the baby was a member of a unit called *Tiger Force*. This was a platoon consisting of forty five men organized out of the 1st Battalion, 327th Infantry, as a long range reconnaissance patrol, to be the eyes and ears of the battalion commander. Its mission was reconnaissance but soon warped into a mission of destruction and the murder of hundreds of Vietnamese civilians. The name of this soldier was Private Samuel Ybarra, a San Carlos Apache from Arizona.

By this time I had been with the Heilbronn Resident Agency, Second Region USACIDC, for nearly half a year. And although some men of this unit were now serving in Germany, we were not tasked to seek and interview any of them.

When this CID investigation ended in 1975 it had been ascertained that many of the men in *Tiger Force* committed war atrocities and that the commanders of the unit, including the battalion commander, knew about them but chose to ignore the crimes. Over a seven month period in 1967 it is believed this LRRP[38] unit murdered hundreds of civilians and mutilated their bodies by scalping and the taking of ears. Yabarra, identified as the worst offender was out of the Army by the time of the investigation and outside the CID's jurisdiction. When the 55 page investigation was presented to Army authorities it was decided not to prosecute anyone.

Years later the atrocities were investigated by journalists Michael Sallah and Mitch Weiss who wrote a book titled *Tiger Force* published by Little, Brown and Company in May of 2006.

During the year I was with the agency I helped investigate one murder, one suicide, and a number of thefts, assaults and robberies. But by far the most crimes I investigated involved the use or distribution of illicit drugs. There was an incident just before I joined the agency that is worth mentioning. Later it was rather amusing, but at the time it caused me some worry.

The week before my transfer I was tasked with the additional duty as the battalion CQ or Charge of Quarters. I was to remain awake all night at a desk just inside the headquarters entrance along with my CQ runner, a junior enlisted man. I was to challenge and identify anyone entering the headquarters. If required I was to call the duty officer for assistance.

---

[38] Long Range Reconnaissance Patrol.

Later that evening two CID agents from the Heilbronn office came in and identified themselves stating they wanted to ask me some questions. Assuming they were there to interview me before my arrival I said sure. But their questions soon alerted me that they were not there for that reason.

The first thing they asked was where I had been on the evening of the previous Saturday. I paused, wondering what this was about. I replied that I had been in Heidelberg, watching the lighting of the castle, an event held there each fall. They asked me if anyone could verify that. I replied that yes, my wife, and my friends Carol and Rick Palm, as well as Rick's two sisters Cathy and Karen Palm, had all been there with me. They seemed to relax a little but the questioning went on.

What time did we get home, did I have any missing laundry with my name on it. I replied that we had gotten home late because the lighting ceremony was a night time event. And I didn't know if I had any laundry missing. I finally lost my patience and asked them what it was about.

They explained that on that same Saturday night a German woman had been dragged into my storage room in the basement of my building, laid down on a towel with my name on it, and raped all that night and most of the next day!

I was shocked to say the least. I explained that since Carol and I had nothing stored in the basement storage room it wasn't locked. I stated that as they knew, washed laundry was hung out to dry on clotheslines in the attic section of each residential building, so anyone could have snagged my towel. They seemed satisfied with my answers and left.

I later found out that the woman in question was the beer man's wife. This man delivered beer by the case and keg to our housing area. I also found out that the woman had attended a party in my building, left with another German national, picked up my towel, and they spent the weekend partying in my basement storage room!

But, it was not a good way to get introduced to the men I would be working with in the CID for the coming year.

When I decided, after a year of working there, that the CID was not what I wanted to do, the chief warrant officer, Edward R. Thomas, wrote me a glowing evaluation with "outstanding" in all areas and said:

"Sergeant Cross has performed the duties of a CID Agent. He has exhibited maturity, versatility and a high degree of professionalism in every aspect of his job performance. He works well under pressure, always displaying a sense of good humor and the ability to resolve complex situations at a given moment. He possesses first sergeant potential and is well oriented."

Mr. Cross, CID agent-in-training of the Heilbronn Agency, pocket protector and all. Yes, it was a time of the wide tie.

CERTIFICATE OF TRAINING

This certificate attests the completion by

WALTER L. CROSS

of a Project TRANSITION training course in

LAW ENFORCEMENT

during the period from 10 August 1970 to 18 September 1970

Department of Defense Project TRANSITION provides training in skills needed to enter civilian employment following military service.

240 HOURS OF LAW ENFORCEMENT INSTRUCTION FROM SPECIALLY SELECTED SUBJECTS, INTERNATIONAL ASSOCIATION OF CHIEFS OF POLICE, (IACP), PROFESSIONAL STANDARDS DIVISION, WASHINGTON D.C.

CROSS SUCCESSFULLY COMPLETED THE PRESCRIBED FBI RANGE FIRING COURSE WITH THE CALIBER 38, POLICE SERVICE REVOLVER AT FORT CAMPBELL, KENTUCKY, TRANSITION PROGRAM. CROSS ACHIEVED A SCORE OF 70.

Murrey E Morris
SGM MPC (Ret)
Instructor

Merlyn V. Burton
Co-ordinator Project Transition

3AA Form 336
1 Nov 68

My earlier training in law enforcement at Fort Campbell was one of the reasons I was assigned to the CID Agency in Heilbronn.

BOOMER SOLDIER

Chapter XI
The U.S. Army Recruiting Command

By this time Jerry Shaw was firmly ensconced as the battalion legal clerk, Company A had a new company clerk, and the battalion still had an assigned career counselor. I was now excess to the battalion. I needed to figure out what route I wanted to take in the Army. So I worked with the career counselor and put in a request to be an Army recruiter.

I was surprised when within a few short weeks an interview team from the United States Army Recruiting Command came to Wharton Barracks, and I was interviewed for possible selection as an Army recruiter.

The interview took place in the conference room at the battalion headquarters. It was similar to a promotion board and was comprised of senior NCOs and a few officers. They asked me fairly simple current affairs questions, my opinion of Army and Defense Department policies and about Vietnam.

The last question was asked by a master sergeant.

"I see you have two Bronze Star Medals for valor and meritorious service. How did you get two of them within thirty days of one another?"

He was smiling, and the question was rhetorical, I simply chuckled. They told me right then I was selected for recruiting duty. They told me I would receive a six month drop from my overseas tour and that I should get my affairs in order because the orders would come soon. And they did. In fact my European tour would be terminated immediately.

My selection for recruiting duty was not the only thing I was happy about. The month before my selection in July of 1973 I had completed my studies with the University of Maryland through the post education center and earned my Associate in Arts degree. My German woman education counselor was pleased too, I was the first of her assigned students at Wharton Barracks to complete a degree. It was the first of three degrees I would earn thanks to the U.S. Army and the GI Bill.

Chicago District Recruiting Command

After a thirty day leave home to visit my folks in Oklahoma and Carol's family in Texas, as well as attending the Army Recruiter School, I reported to the U.S. Army Recruiting Station in Des Plaines, Illinois.

Des Plaines, pronounced the way it's spelled, to include sounding the two letters of 's' is a northern suburb of the City of Chicago. Carol and I immediately moved into a government leased apartment at 515 South Milwaukee Avenue in the next town north of Des Plaines, the town of Wheeling, Illinois.

Wheeling, more so than Des Plaines, was a working class suburb and became a part of my assigned recruiting area. Our furniture soon arrived from Germany, despite missing some items. After settling with the transportation office for the lost goods, Carol and I bought new furniture for our first ever civilian apartment as a couple.

After graduation from the U.S. Army Recruiting School I was assigned to the Chicago District Recruiting Command or DRC. This is a publicity photo taken by the United States Army Recruiting Command or USAREC in September of 1973 that was eventually made into a recruiting poster, I was 24 years old.

Carol had a lot of fun meeting our new neighbors which included the apartment manager, a lady from Germany who married a World War II veteran and her teenage daughter, Andrea, who went by the nick name of Cookie. Soon, Carol was enjoying the swimming pool with Cookie and exchanging recipes with her mom. We liked the area, although I could tell it was going to be cold, but I had no idea yet just how cold it would get.

I reported to the recruiting station at 800 Lee Street in Des Plaines and found it to be a four man station with three recruiters and a station commander who was also "on production". In other words he had to recruit volunteers as well as supervise and be responsible for the recruiting station, its equipment and vehicles.

The station was co-located with the other three services, the Navy, the Air Force, and the Marines.

When I got out of the Army for those eleven days back in October of 1970 I reenlisted in Houston. I listed the address of Carol's mom and dad as my home of record because that is where we were living at the time.

In December I got my first assigned recruiting mission and promptly enlisted a high school diploma

This picture appeared December 20, 1973 in the Wheeling Reminder newspaper under the headline "Wheeling has a New Army Recruiter". The caption reads: "Sgt. Walter L. Cross is the new Army Recruiter for the Wheeling area. Sgt. Cross is from Houston, Texas and entered the US Army in 1967. Sgt. Cross served in Vietnam during 1969 and 1970. He was stationed in Germany prior to his assignment to the US Army Recruiting Station, 800 Lee St., Des Plaines. Sgt. Cross resides in Wheeling with his wife, Carol.

graduate[39] from Wheeling High School named John V. Nystrom. John enlisted for mechanic school. Later, I heard a rumor that this man, the first recruit I put in the Army, was discharged after ninety days.

---

[39] Our primary group for recruitment was high school graduates.

John Nystrom heads for Army Mechanic School

This picture also appeared in the Wheeling Reminder, the individual pictured with me is John Nystrom, the first man I enlisted into the U.S. Army. The caption reads: "John V. Nystrom, son of Mr. and Mrs. Vern Nystrom of Wheeling, enlisted in the Army for three years. John entered active duty on Feb. 12th, 1974 (I had actually enlisted him in December, 1973). John enlisted for the choice of training, choice of station option. Upon completion of mechanic school, John will be assigned to Fort Carson, Colorado. Young men and women interested in this option or any of the several options offered by today's Army, are invited to visit the Des Plaines Recruiting Station on 800 Lee Street.

BOOMER SOLDIER

On February 8, three Wheeling men enlisted in the Army's delayed entry program. They are: Bob Richardson, 1200 E. Lee Street, Wheeling, son of Mr. and Mrs. Bob Richardson, Wheeling. Bill Barron, son of Mr. and Mrs. William Barron, 1200 E. Lee Street, Wheeling. And Bill Anfeldt, son of Mr. and Mrs. Ernest Anfeldt, 873 Fletcher Drive, Wheeling. All three of the men enlisted for Heavy Equipment Mechanic School and assignment to Germany. The men will enter active duty on the 20th of March, 1974. This is only one of over three hundred jobs offered by Today's Army. For further information on the jobs offered by the Army, contact Sgt. Walt Cross at 824-0821 or visit the Army Recruiting Station at 800 Lee Street, Des Plaines.

The particulars were rather vague so I'm not sure why he was discharged.

John's enlistment led to the enlistment of three of his friends, and I was well and truly on my way to becoming an Army recruiter. The enlistment of these four individuals plus a couple of others, led directly to my meritorious promotion to staff sergeant.

These three new soldiers stuck, and to my knowledge served their full three year enlistments in Germany and received honorable discharges.

It took only a few months of exceptional recruiting performance before I was meritoriously promoted (no promotion board required) to staff sergeant in February, 1974. Although that is Major Douglas, the area commander, who is shaking my hand, it was Captain Mayer who is standing behind me, the deputy area commander that recommended me and pushed for my promotion. Both these officers are Vietnam veterans.

I believe that the enlistment of these four men led directly to my immediate promotion. The recruiting area commander, Captain Phillip W. Mayer recommended me for meritorious promotion to staff sergeant writing:

"Sergeant Cross has performed the duties of a field recruiter in an excellent manner. He is a conscientious individual who has constantly given the recruiting area his total support and loyalty. Sergeant Cross identifies well with the younger generation, and creates a favorable impression with those he comes in contact with. He participates in the equal opportunity program, and with experience will develop first sergeant potential. He should be immediately promoted to the rank of staff sergeant."

I had been a recruiter on full production status for two months! At first blush recruiting duty seemed to agree with me. But, first impressions can sometimes be misleading.

I recall that during the four years I served in the Des Plaines recruiting station, I had five different station commanders. I can recall all five of them but of the five, only three of their names.

The first I recall was an interim station commander assigned to us until a permanent commander was selected. I don't remember his name, but he was a sergeant first class of the old school. Well in his late forties, he sported a crew cut of grey and blond hair and had a very aggressive attitude. I never learned anything of his back ground. He did not last long as he was discovered molesting a sixteen year old girl working for the station as an office aid. I don't know what became of him other than the fact I know he was court-martialed at Fort Sheridan.

The second was Sergeant First Class Jerry Klemme. Jerry had been a staff sergeant with me until his recent promotion. He was married to an Asian woman he met while on a tour of duty in Korea. While still just a recruiter he enlisted a fellow veteran named William Peters from my recruiting area of Wheeling. I still don't remember how he got a man from my area; perhaps I never knew about it until later. Taking from another recruiter's area was known as poaching. So right away I viewed Jerry with suspicion.

Billy, as we came to know him, was a Vietnam veteran and went in at an advanced rank. He would soon return as a recruiter and become a very good friend of mine, not to mention a good recruiter. He spent the remainder of his career on recruiting duty and retired a master sergeant.

At first Jerry was pretty good to work with as he understood the pressures of recruiting duty. But later he became abusive and would demand any offending recruiter to "Get in the back room!" There he would proceed to berate them loud and long. This bothered me no end, but since he was the station commander, there wasn't a lot I could do about it. He always seemed to avoid me, perhaps because I was the next recruiter in experience if not date of rank, to him. Then one day he singled me out with his trademark yell of "Get in the back room!" I was not about to be intimidated by this man.

I stood up from my desk and picked up a heavy standing ashtray and replied.

"Fine, but one of us is not coming out!"

Caught off guard and completely frustrated, Jerry stormed out of the office. He never accosted me again, but his abuse of the other junior NCOs continued and even grew.

Finally, one day when Jerry was out of the office, the other recruiters approached me and told me "You gotta do something about Jerry". I replied by asking why me? They went on about how I was the most senior recruiter, probably not true, but I accepted it. I told them to get into the Army car we were going to go see Captain Mayer, now the Recruiting Area commander.

When we arrived I told them to follow me and do exactly what I did. I walked into the captain's office taking off my recruiting badge as did the other recruiters. I stopped and saluted and laid my badge on his desk and the others followed suit. He frowned at us and spoke to me.

"Staff Sergeant Cross, what's the meaning of this?"

I outlined what we were experiencing with Jerry as station commander and the others joined in to confirm what I was saying. I summed up by telling him we felt so strongly that it had come down to either Klemme left, or we all would. He leaned back in his chair and studied me a moment and then said.

"Go on back to the station he'll be gone when you get there."

And when we arrived, he was nowhere to be seen. Later we learned he was sent to another recruiting station, but only as a recruiter. I would see him one more time, when he passed through my office on his way overseas. But by that time I was off of recruiting duty and working as the senior personnel sergeant at Fort Sheridan.

Wheeling friends, the girl to my left is Vicky Bjornson a longtime friend that kept in touch with us for many years and still does. She was a very amusing person with a great sense of humor. The other woman was known as "Big Vickie". They lived in our apartment complex.

The next station commander was another sergeant first class assigned as temporary station commander. I don't recall his name, but I do remember he was a martial arts expert. He would play a role in an unusual recruit I enlisted.

One particular day I was manning the station while the station commander and the other recruiters were out. They were busy taking recruits to the AFEES[40] in downtown Chicago. I took the chance to catch up on my paperwork. The office and indeed the entire building were quiet. Not even the Marines were laughing and jibing one another telling their war stories. My head was down concentrating on what I was doing. Suddenly, a yell ripped through the air, like in a Pink Panther movie when Kato would attack Inspector Clouseau to keep his defense skills sharp.

"Hhhaaaiii Uppp!"

My head snapped up just in time to see an Asian man hurtling through the air, his hands up in the classic karate defense posture. He landed flat footed in front of my desk and yelled

"Me go Korea!"

I looked at him, kind of stunned and managed to say "Sure, uh, sure, have a seat Mr.?"

He straightened up and bowed "Kyu Se Pak!" He informed me.

I put Mr. Kyu in the U.S. Army, I sent him to the Army food inspection school, but he didn't seem to care all that much what he did as long as he got to go home to Korea. His father surprised me later during the talk we had before Pak was processed for enlistment.

His father owned a Korean food restaurant and I traveled with Pak to speak to his father. The talk

---

[40] Armed Forces Entrance and Examination Station

went well and when the decision was made for him to enlist, accompanied by giggles from the Korean girls who worked in the restaurant, his father insisted I have lunch. I ordered and enjoyed some very good Korean barbequed steak. And then his dad pulled the surprise. He slid a fat envelope across the dinner table to me.

"What is this?" I asked, but suspecting what it likely contained.

"For you Sergeant-san, you most welcome." He replied.

I opened the envelope gingerly and noted a very large stack of twenty dollar bills. I explained to him that in our culture it was not necessary to pay a government agent for services. He took a while to convince, and wasn't happy, but finally accepted the fact I was not going to take his money for enlisting his son.

As one of the few recruiters who managed to spend my entire tour in the Des Plaines station, I was often left as the only man in the station. I heard we were going to get a new station commander and if memory serves I seem to recall it was in the spring of 1976 that I met him.

Working on paperwork at my desk I was searching for something in a bottom desk drawer. When I sat back up in my chair I exclaimed "What the hell!", because I was looking at a pair of highly shined low quarters[41] planted firmly in the middle of my desk standing on top of my weekly planner.

I slid my chair back and looked up to see a short stocky sergeant first class with a neatly combed over dark head of hair, wearing Vietnam service ribbons, a combat infantryman's badge, and the blue infantry shoulder cord, standing with his fists on his hips. In answer to my query he said.

"Greetings, my name is Tom Baldwin, I am your new station commander."

And that is how I met Thomas Victor Baldwin. He would prove to be an exceptional station commander and a good friend who went to bat for me on more than one occasion. Like I did Billy Peters, I liked Tom a lot and looked up to him, despite his short stature.

Tom proved to be a good station commander and when our recruiter strength was increased he appointed me his deputy station commander.

---

[41] Army black dress shoes.

Des Plaines assigned Army recruiters: Norbert Zavadin, Walt Cross, Tom Baldwin, Billy Peters, and David Andorka. I'm smiling because I just completed my third and (I thought) final year of recruiting duty. Shortly after this picture appeared in the Wheeling Journal, I was involuntarily extended for an additional year. We weren't happy about it, but Carol and I both soldiered on. Eventually we and the other Army recruiters scattered across the U.S. and even overseas in areas like Guam and Puerto Rico and on major

U.S. bases in Europe, we built the All Volunteer Army and the draft was ended for good.

Kyu Se Pak shakes hands with Staff Sergeant Paul Duval. I was out of the office when Kyu returned from enlisting and Paul stood in for me when the Des Plaines reporter took this photograph.

Off duty: Carol and I lived in a government leased apartment in Mount Prospect at this time. The chair I'm sitting in was my first easy chair, but that is not my first beer. I can't tell what brand of beer I was sipping. Carol and I watched the premier of a brand new show from that chair, called *Saturday Night Live* on October 11, 1975. As you can see I have a note pad for writing close to hand. I'm not wearing a mustache in this picture I can't remember why I shaved it off.

The flight approach to O'Hare Airport was right over our apartment building; Carol and I were often awakened by low-flying jet airliners. Not long after we left Mount Prospect it was the scene of a catastrophic airplane crash.

Carol and I moved into a very nice government leased apartment in Mount Prospect early in 1976.

It was about this time that a young Army specialist, Craig Armstrong, was assigned temporarily to the station as an Army recruiter aid. He was originally from my recruiting area of Wheeling and so he teamed up with me.

He was a good looking young soldier with blue eyes and a shock of blond hair and the high school kids took to him right away. He would prove to be a big help to me.

While he was assigned to us I took him home to have dinner with Carol and I at our apartment in

Wheeling. During his visit we introduced him to Andrea, also known as 'Cookie', the daughter of our land lady, a very pretty young woman with strawberry blond hair and Craig's age. They hit it off right away and it wasn't long before Craig asked me to be his best man at their wedding.

A photograph of Craig Armstrong and I the evening of his wedding to Cookie, the daughter of my landlady. I was his best man and so I wore my dress blues to the ceremony in the summer of 1974. This was many years before the dress blue uniform came to be the standard uniform worn by the Army.

Recently I managed to make contact with Craig and Andrea who returned to Illinois after his tour of duty in Germany ended. We had lost track of them and it wasn't until September of 2018 I found their contact information. Since I don't have a picture of Andrea at their wedding I have included a picture of them forty-four years after they wed.

Through hard work by both of them they have made themselves a good home and raised a family.

Craig is the proverbial 'old Army buddy' you always hear veterans say when referring to those with whom they served in the military, or at least in the Army.

Craig and Andrea Armstrong in 2018.

My recruiting duty tour was supposed to be completed in September of 1976 and I couldn't wait for it to be over. The stress had really taken its toll, my migraine headaches had returned, with a vengeance. On more than one occasion I had to go to the hospital to receive a shot of Demerol, a rather potent pain killer. These episodes continued, intensified, and carried on even after I left recruiting duty.

But there was to be no letup in the stress. I was involuntarily extended on recruiting duty for an additional year. I wasn't happy about it, but proceeded to do my duty, albeit with growing resentment.

Not long after Tom arrived we moved our office from the back of the building to the front in a shuffle ordered by the Department of Defense. It was the Army that had the biggest recruiting mission and needed to be the first service any walk-in recruit would see.

While in the new office we also received our first, and only, woman recruiter during my tour of duty. Her name was Jodi Harvel and she was both professional and attractive in appearance.

Tom was out of the office when she reported for duty and as the deputy station commander; I welcomed her to the Des Plaines station, showed her to a desk and began to give her the rundown on the recruiting area.

About that time the phone rang and excusing myself I answered it in my office.[42] It was Tom and

---

[42] The senior recruiters in rank and experience got their own individual offices within the larger office while the newest and least in rank were

he started the conversation by saying "Is she there?" When I answered in the affirmative he continued "Does she bark?"[43] I answered no and he came back with "Does she whine?" I answered no that she did not and he replied that he would see when he got there.

Carol's brother Lance visited us at Mount Prospect and caught the largest catfish ever taken from the city park pond. Later we took him to the amusement park *Great America* in Gurnee, Illinois.

And of course he did, and was impressed despite his misgivings. Jodi would go on to have success on recruiting duty and was still at the station when I

---

placed at desks in the large outer office. That was where Jodi would start.
[43] A reference to his assertion earlier that the Army would likely send us a *dog* meaning an ugly woman. Which Jodi certainly wasn't.

finally did leave recruiting duty. A couple of amusing incidents happened while I was in the front office.

Sometime earlier I made a suggestion to the area commander that upon our movement to the new office our phone number be changed to 297-ARMY. This was readily done and years later when I happened to visit an Army recruiting station in Stillwater, Oklahoma I noted their number was also 000-ARMY. I'm not sure I can claim that I had the original idea, or that my idea continues to this day, but to my knowledge this was the first use of ARMY for the end of the telephone number of an Army recruiting station.

That number rang in my office and I answered it one day with my usual official greeting of:

"Army recruiting station, Staff Sergeant Cross, how may I help you?"

The voice of a young man replied in an almost desperate tone; "I've got to have the 82nd Airborne Division!"

I hesitated a moment and then asked "Do you want me to have them parachute into your back yard?"

He came back with "No, no, I want to join the 82nd Airborne!"

I asked him his name and then told him to get down to the recruiting office as soon as he could and ask for me. I would put him in the 82nd Airborne!

He did and I did. He enlisted for six years in the infantry with a bonus and was guaranteed airborne training and if he completed it, assignment to the 82nd. That was a fun and easy enlistment, he was a good man and came back all shiny and tucked into his spit shined jump boots. He was indeed a proud soldier and I was proud of enlisting him.

In December of 1976 I received my third award of the Good Conduct Medal. This was the only award besides my recruiter badge and the many letters of commendation I would receive during my four years of recruiting duty. This award was for three years of "Exemplary conduct, efficiency, and fidelity in active Federal military service."

I made an error one day, that would come back to haunt me, and pointed out the dangers inherit in idle talk. About to leave the office, I was standing in the doorway with my back to the hallway. The subject of my conversation with the other recruiters was our deputy area commander, Master Sergeant Thomas Augustus Bragg.

As you will be able to tell I did not care much for Master Sergeant Bragg, I felt he was arrogant, knew little about the art of recruiting new soldiers, and had hurt the careers of many recruiters, including some I considered friends. In response to a remark from one of my office mates I said "Master Sergeant Bragg is a back stabbing bastard!"

And then, looking over my shoulder, who was standing there but Master Sergeant Bragg! I simply moved to one side of the doorway allowing him entrance. I did not apologize nor did he say a word. Later it occurred to me he likely thought that I knew he was there and wanted him to know, indirectly how I felt about him.

Needless to say my efficiency report upon my departure was not the highest I ever got. But, to his credit he gave me the highest rating in the leadership skills area that read "Staff Sergeant Cross is clear and to the point in conveying information…" how ironic.

What was not written down was the statement Bragg made when signing my EER.

"You have been selected to attend the advanced NCO school, which means someone wants you to go to the Sergeants Major Academy. But that won't happen with this EER."

Well, he was quite wrong. I got high marks at the advanced school and when I became eligible, was indeed selected for attendance at the academy. However, that EER likely kept me from being promoted to sergeant first class E-7 the first time I was considered. I would have to wait at least until the second round before I was selected.

These actions however were still in the future. I continued to serve, now involuntarily, on recruiting duty for the time being.

A couple of months before leaving recruiting I paid a visit to the Glenview, Illinois Naval Air Station to complete paperwork for a Navy reservist to enlist in the Army.

Naval Air Station Glenview supported the Naval Air Reserve, Marine Air Reserve, the 4th Marine Aircraft Wing, and the U.S. Army Reserve's 244th Aviation Group as well as an active duty Coast Guard Air Station.

While there I spoke to the administrator who was in civilian clothes and asked him about his job. He explained that he was a Department of Defense employee but was required to be a member of the Navy Reserve. It sounded like a good position to be in as he drew his salary, his reserve pay, and earned time toward a military retirement at age sixty-two. He explained that all reserve forces, including the Army Reserve had similar positions.

I stored the information away, and told Carol about it, not knowing that it would pay dividends in the future.

Finally in October of 1977 I received orders to report to Fort Sheridan, Illinois just a few miles down the road from the Des Plaines recruiting station.

I remain in touch with both William (Billy) Peters and Tom Baldwin. All three of us retired as Army master sergeants.

Billy Peters and I doing some recruiting in a Chicago mall on the 4th of July during America's Bicentennial year of 1976. The Army provided these Continental soldier uniforms for this special event. People signed a copy of the Declaration of Independence at our booth and the document was forwarded to the National Archives.

Chapter XII
Garrison Duty - Fort Sheridan, Illinois

Non Commissioned Officer in Charge

SSG W. L. CROSS (P) NCOIC

My desk nameplate at Fort Sheridan.

You may recall that I served as a personnel specialist after returning from Vietnam. And when I left recruiting duty behind I still held that job designation as a secondary MOS. Except now I was a staff sergeant and so I was designated a senior personnel sergeant.

It just so happened that when I came up available for reassignment, Fort Sheridan, where the U.S. Army Recruiting Command was headquartered, needed a personnel sergeant. And so I was assigned just thirteen miles down the road from my recruiting station, to Headquarters, U.S. Army Garrison, Fort Sheridan, Illinois.

I was now the NCOIC (Non Commission Officer in Charge) of reassigning, promoting, reclassifying and other personnel actions for assigned Army recruiting and base personnel.[44] I was authorized on base housing, and effective October 26, 1977 I reported to the Personnel Division at Fort Sheridan.

Carol and I moved into our new home on base on a Friday. It was a well maintained, white, two story duplex that backed up directly to Lake Michigan. Our back yard literally ran about twenty yards to the east of our back door and then dropped precipitously to the rocky shore of the huge lake.

The lake itself is so wide that the other side is, of course, invisible from our side. We lived in the north end unit of our duplex which ran north and south, and we liked it very much. After a full day of moving we were tuckered out, and were grateful we had a weekend to recover and could sleep in Saturday morning. Not so fast!

There we were, snug in our warm bed, safe from the cold October morning and sleeping like babies. All of a sudden a moose was standing over our bed and bellowing. No, not really, but that was what it sounded like. I jumped outta bed and looked out our second story bedroom window. I could just make out some dimly seen hulking object in our back yard. Things were dimly seen because there was a thick fog completely obscuring the lake and most of everything else. Then the noise sounded again, and I

---

[44] Officially NCOIC Promotions/Reclassification Section, Personnel Management Branch, Military Personnel Office (MILPO).

realized it was a fog horn. What I was seeing was a monstrous tanker that had pulled up along the shore due to the fog and was warning other craft away with its fog horn. That was a strange experience, but only the first we would endure as we learned Lake Michigan is so big, it generates its own weather.

I got back on my college degree program, set aside for the four years I'd spent on recruiting. By the time I was ready to leave Fort Sheridan I had completed my studies at the State University of New York or SUNY, and earned my bachelor's degree in history, all paid for by the U.S. Army.

At the Personnel Division, often referred to simply as Personnel, I worked directly for Sergeant First Class Cooper, who reported to the chief warrant officer. There was a third NCO, Staff Sergeant Steve Woznick who handled administration and I, as I mentioned, supervised all personnel actions.

There were a number of clerks both men and women who worked in personnel, as well as a few civilian employees. The civilians only answered to the chief.

All female soldiers were members of the Women's Army Corps or WAC which had not yet been disbanded.[45] The Greek goddess Pallas Athene was the official symbol of the WAC.

One of our daily duties was the cleaning up of the grounds of our building and in military jargon is known as police call. This includes the picking up of trash including things as small a cigarette butts. During this duty the NCO in charge followed along behind to make sure nothing was missed by the detail. Sometimes I had to call one or two of them back to pick something up they missed. As a joke, and I always suspected it was one of the WACs, someone took a pink plastic baseball bat and with a black marker wrote on it "WAC Whacker" and left it on my desk. So to continue the joke I used to carry it during police call. Yes it was politically and probably militarily incorrect, but we all got a kick out of it. I always suspected it was Specialist Martha Cheatham that had made this object for me. Martha was a very pretty light complexioned black woman in her mid-twenties and tended to be mischievous.

---

[45] The Women's Army Corps was disbanded by an act of Congress on October 29, 1978.

She later married an Army captain (also a doctor of Veterinary Medicine) of the Veterinary Corps.[46] There were other personnel specialists working for me and Wozniak but I can't recall their names.

On occasion when 'the chief', our nickname for the personnel section warrant officer, would leave the office for some reason I would stand up and say in a loud voice "Attitude check!"

Everyone would pause in their various tasks and we would all sing the familiar phrase we knew so well.

"We like it here, we like it here, you fucking 'A' we like it here! We wanna stay, we wanna stay, you bet your ass we wanna stay!"

Then we'd all laugh or at least chuckle and go back to work with a smile on our face and a little lighter feeling in our hearts.

We were not aware of it, but we were in the last days of the Women's Army Corps, and it was disbanded the very next year. The official demise of the corps really had no effect on our day to day duties.

One day, not too long after I assumed my new duties I picked up the morning distribution and came across a familiar name. Sergeant First Class Jerry

---

[46] Most Army veterinarians don't treat animals, they inspect food.

Klemme, my old station commander, was up for reassignment from recruiting duty.

He had filled out his "dream sheet" or request for where he would like to be assigned and indicated his preference for an assignment to the Republic of Korea. I did my due diligence and checked for any need in Korea for a soldier of his pay grade and military job description and unfortunately for him, found none. Then I checked Europe for an opening and I must admit, I was happy that I did find one, several in fact. I soon had orders cut assigning Jerry Klemme to a three year tour of Germany. I expected some reaction and it was not long in coming.

Sergeant First Class Klemme came stomping into the personnel office a few days later demanding to know who had sent him orders to Germany when he had requested an assignment on the other side of the world in Korea!

One of my clerks pointed toward my desk. Jerry looked over at me and his eyes opened wide in surprise as he first looked at my face and then glanced down to the name plate on my desk that read: "Staff Sergeant Cross NCOIC". He mumbled something under his breath that sounded suspiciously like "I'll be a sonofabitch!" or maybe he called me a sonofabitch, and turned toward the door.

"Hey Jerry, hold on, I was gonna ask you out to lunch." I said, but he never slowed down or looked back. And he never applied for concurrent travel of dependents to Europe, so I assume his Korean wife didn't want to go with him. This is a classic case of the old refrain to watch who you butt heads with, because what goes down does indeed come around.

One of the more unusual levies or requests for personnel was for six month temporary duty assignment (TDY) of supply sergeants to, of all places, Eniwetok (or Enewetok) Atoll in the South Pacific.

Eniwetok is a large coral atoll in the Marshall Islands and was the scene of bitter fighting between the U.S. Marines and the Imperial Japanese Army during WWII. The land surface is less than two and half square miles in size and was inhabited by Polynesian Eniwetokians before the war.

From 1977 to 1980 the U.S. Army constructed a massive concrete dome to enclose radioactive debris collected from the island so that eventually, 850 former residents could reoccupy their traditional homeland. The supply sergeants I sent were to assist in this humanitarian effort.

This work was to restore Eniwetok to its former condition prior to being used as a nuclear target by the U.S. no less than forty three times from 1948 to

1958. The island was opened for habitation in 1980 after costing U.S. tax payers millions of dollars.

An American fighter bomber strafes the enemy hidden in trenches as Marines lie prone in the sand during the battle for Eniwetok Atoll.

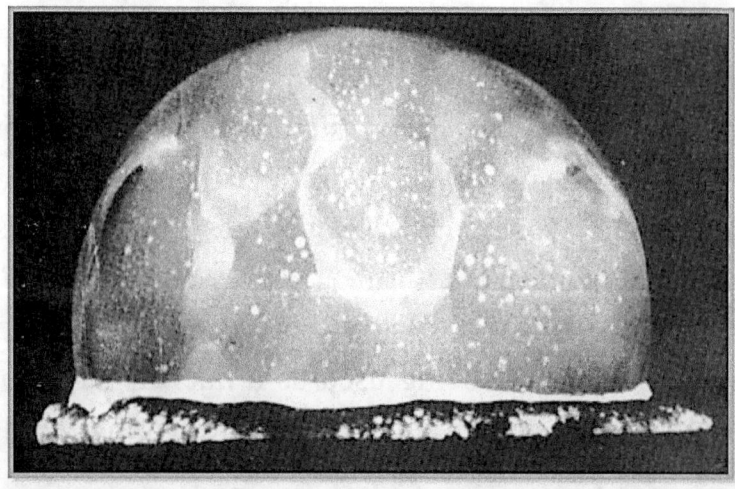

One of the forty-three nuclear bomb detonations on Eniwetok Atoll during the decade of 1948 to 1958.

This is the Runit Dome on Eniwetok Atoll. Nuclear waste is stored beneath the dome. The dome is placed in the crater created by the "Cactus" nuclear weapons test in 1958. I sent at least a half dozen supply sergeants there to assist in this humanitarian expedition.

BOOMER SOLDIER

In January of 1977 President Gerald R. Ford created the Humanitarian Service Medal. I don't know if any of the men I sent to Eniwetok ever received this medal, but if not, they should have, in my opinion. There was of course, some personal hazard working around radioactive soil and debris.

One last note about this unusual assignment is that I always met resistance when I selected a soldier to go TDY to Eniwetok. All but once, when I had a sergeant call me specifically and volunteer. I sent him, but his tour was terminated and he was brought back to Fort Sheridan under arrest. It seems that before he left, he stole and sold a rather large amount of Army equipment and supplies. I'm not

sure what he was thinking about when he fled to an island. There was nowhere for him to go, or to hide.

In January of 1978 I went on temporary duty to Fort Benjamin Harrison, Indiana to attend the Non Commissioned Officer Advanced Course. I was there from 4th of January to the 3rd of March. This advanced course, which I completed well toward the top of the class, led later to my acceptance at the U.S. Army Sergeants Major Academy. The class consisted of 47 staff sergeants, four of them women.

I am center left in this section of the class photograph.

One of the extra duties that rotated to NCOs of the Fort Sheridan garrison was that of NCOIC of the post color guard. This duty came around to me in July of 1978. I performed the duty, drilling the color guard and ensuring they understood their duties as it pertained to attending military funerals, military retirements, and other types of ceremonies. After the end of my thirty day stint of duty I received a letter of appreciation from the garrison commander that reads as follows:

1. I wish to take this opportunity to officially express my appreciation to you personally for your outstanding performance as NCOIC of the United States Army Garrison Color Guard for the month of July 1978.
2. Your personal management of the color guard missions and the people involved was excellent. Your concern for the mission and your people during and after the commitment month was excellent.
3. I offer a heartfelt thank you to a professional soldier.

> Gary R. Delozier
> Captain, Infantry
> Commanding

United States Army Garrison Color Guard – Fort Sheridan.

I personally found the assignment challenging, interesting, and an honor to perform.

On the home front during that same year Carol and I got to experience more of Lake Michigan's unusual weather. At the end of 1977 we had 37 inches of snow already on the ground. On the second day of 1978, it snowed another 36 inches of what was known by the weathermen as "lake effect" snow!

We woke up to snow over our front door; we could not get out of our quarters! That day was declared a training holiday and all but essential personnel were released from duty. I found that if I went out the back door I could slide along the house siding and reach the north corner of the house. That is where I began to shovel, and shovel, and shovel, and shovel some more until I cleared a path around the house to the front door. Then, I shoveled out our car. The only good thing that day was I found a twenty dollar bill in a snow drift. The next day the engineers had cleared the parking lot and roads and we all were able to return to duty.

During 1978 I was under consideration for promotion to sergeant first class by a promotion board in St. Louis. I had hoped that my completion of the advanced NCO course might be enough to overcome the negative remarks of Master Sergeant Bragg a year before. But I knew my chances were pretty slim, and when I was passed over for promotion I decided, after talking with Carol, that it was time to leave the Army. But I had another year to serve on my current enlistment. With the decision made Carol and I settled into domesticate tranquility on the banks of that mighty body of water, Lake Michigan.

One of my fondest memories of the time is getting to see the Chicago Bears play the Tampa Bay Buccaneers in person at Chicago's Soldier Field. It was an extremely cold day, but running back Walter Payton, nicknamed 'Sweetness' had his usual outstanding game and the Bears won. I remain a fan

This picture of Carol was taken in the backyard of our quarters at Fort Sheridan in 1978. That is Lake Michigan in the background. She was pregnant with twin boys at that time. Sadly, the boys didn't come to term. It was a devastating loss to both us.

of the Bears today although the Dallas Cowboys will always be my first NFL team. I followed Oklahoma University while on active duty, I would not truly discover the joys of college football until Carol and I moved to Stillwater, Oklahoma the home of the Oklahoma State Cowboys. But that was still in the future.

In January of 1979 I put in a request for leave in conjunction with my departure (ETS)[47] from the Regular Army. This is commonly referred to as 'terminal leave'. Upon my departure from Fort Sheridan I would have 45 days leave and I continued to be paid by the Army for that time period.

But before we could leave, a major disaster happened that May, very near our former apartment in Mount Prospect.

A regularly scheduled passenger flight from O'Hare International Airport was designated American Airlines Flight 191. The flight was from Chicago to Los Angeles and took off on May 25, 1979. The airplane was a McDonnell Douglas DC-10 and it crashed just moments after takeoff on the respective borders of Mount Prospect and Des Plaines right in the middle of my former recruiting area. The airplane exploded upon impact and all 258 passengers and 13 crew members were killed along

---

[47] Elapsed Time in Service.

with two other people on the ground. It remains the deadliest aviation accident to occur on U.S. soil. Only a dozen bodies were found intact afterward. The entire community in and around Chicago was in shock. We all mourned the great and terrible loss of life.

I continued my studies through the University of the State of New York extension program at Fort Sheridan's education center. That spring I completed all requirements and earned my bachelors degree in history. It had been a long time coming. Armed with my degree and my many years of military experience, I felt I was ready to tackle the civilian world.

With my terminal leave approved, Carol and I began our preparation for transition to civilian life, and in June of that year we left for Texas to spend some time with her folks in Missouri City. We had a good time but didn't want to live in the Houston area. Eventually we decided to relocate to the small college town of Stillwater, Oklahoma. Stillwater was the home of Oklahoma State University, and there was both a number of Army Reserve commands there as well as an Oklahoma National Guard battalion. I could join up with one of them and use my G.I. Bill to go to Oklahoma State University for a masters degree. As it turned out I did that, and Carol later earned her bachelors degree there.

A few years after I left Fort Sheridan, that post on Lake Michigan with its beautiful white stone buildings was closed as an Army post. It was sold at auction to the highest bidders.

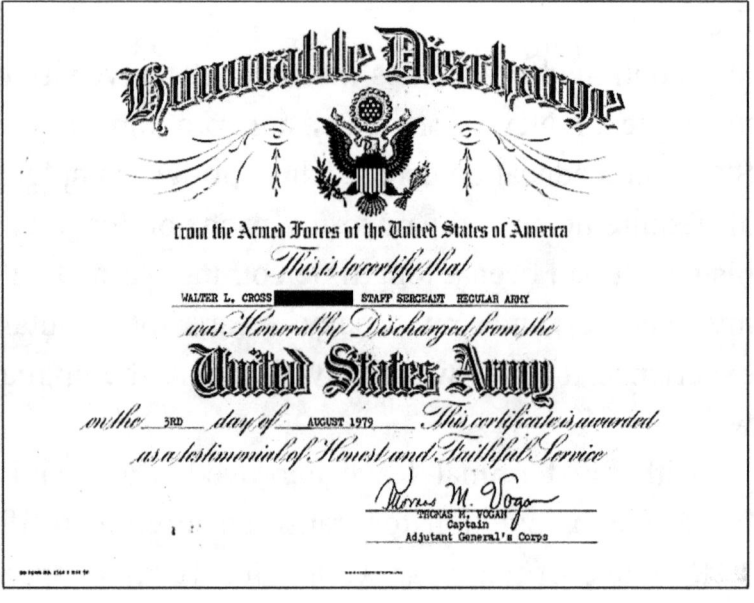

This Honorable Discharge in 1979 ended my twelve years of active duty service in the Regular Army.

Chapter XIII
1st Battalion 291st Infantry Regiment

The Regiment's First Active Guard Reserve (AGR) Soldier

Carol and I found a tidy and comfortable little house in Stillwater at 319 S. Arrington and I began to look for a job. Carol found one first and became an inspector at the Mercury Marine factory on the east side of town. The factory made Mercruiser boat engines and a souped up engine for Chevrolet's Corvette. The only drawback was that when Carol came home at night she smelled like an automobile mechanic!

I didn't have much luck until one day we were driving past the Cleo O. Payne U.S. Army Reserve Center on Washington Street. Carol reminded me of what I had learned at the Glenview Naval Air

Station regarding civilian employees. She suggested that I check with the center about a job. I said that I thought those kind of jobs were kept 'in house' and not usually opened up for outsiders. I did stop by soon after that and got a real surprise. Carol's suggestion would have a great impact on our lives!

I entered the Reserve Center and made contact with three civilians sitting around a conference table having lunch. I soon learned they were Kendal Johnson, the head technician and two other technicians named Guy Short and Johnny Herring. They were very friendly and I explained I was a staff sergeant in the Regular Army on terminal leave. I asked if they had any civilian positions open. Kendal said no, they didn't have any civilian positions, but they did have a military position. I asked him what he meant by a military position.

To my surprise and elation he said they were starting a new program called Active Guard Reserve or AGR where a reservist would be placed on active duty to serve as a unit technician or administrator. I told him I had one week left in the Army and did he mean I could be on active duty and live in Stillwater?

He replied yes and handed me an application. I sat down at a typewriter, filled it out right there, and gave it to him. Then I followed his instructions to go see the Army Reserve recruiter down on Main Street.

In a short time I received my discharge, reenlisted in the US Army Reserve in September and was ordered to active duty on the first day of October, 1979. I was the very first AGR soldier for the 1st Battalion, 291st Infantry Regiment, and for the 3rd Brigade (OSUT), 95th Division (Training). I was also one of, or possibly the very first AGR soldiers, in the 95th Division. My office was in that same Army Reserve Center. It was a match, as they say, made in heaven thanks to Carol's timely reminder.

Ironically, I got back on active duty just in time to receive my first Army Commendation Medal. It arrived in the mail from Fort Sheridan. My four hard duty years on recruiting duty had produced little in recognition outside a meritorious promotion, although I do admit that it was very good recognition indeed! But my eighteen months as a senior personnel sergeant had gotten me my first new personal decoration since Vietnam ten years before. The medal was presented to me by the 1st Battalion commander, Lieutenant Colonel Dwight Stoddard.

Colonel Stoddard would prove to be a very good friend to me, and I have always appreciated him.

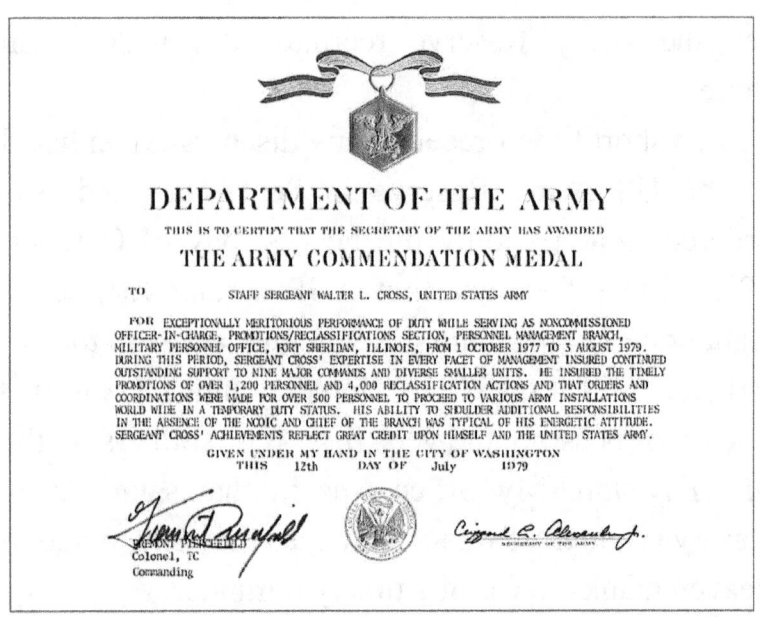

The day I reported for duty I met civilian technician Richard Lee Smith, he would become a close friend and eventually became an AGR soldier as well. Not long after coming onboard I was sent to the Unit Technician Basic Course at Fort McCoy, Wisconsin.

The plane I took from Oklahoma City, took me right back to O'Hare International Airport, just a short distance from my former recruiting station in Des Plaines. There I caught a Republic Airlines prop driven plane on to Fort McCoy. The old plane popped and snapped like it was likely to give it all up at any time. But it did manage to hold together and I arrived for my first visit to the great state of Wisconsin.

I graduated second in a class of nearly 100 technicians, scoring one tenth of a point behind the honor graduate! That really irked me, but he won it fair I assume, and I congratulated him! (But I still like to think the instructor was his buddy).

In the meantime Carol and I purchased a home on Liberty Avenue in Stillwater. Years later someone commented to me that living on Liberty Avenue was an appropriate address to for a military man. To add to our joy, our son Justin was born in May of 1980. He was just a week old when we moved into our new home. It was a happy time for us and Justin was greatly celebrated! He's grown now, has his own children, and has his own home in the nearby small town of Cushing.

Colonel Stoddard had the battalion chaplain christen his 'godson' Justin in the reserve center during a weekend drill. The battalion presented him with a stuffed tiger, the battalion's unofficial mascot because of its affiliation with the Tiger Land training center at Fort Polk, Louisiana. Justin enjoyed that tiger for many years. I think it's still in the top of the closet in his old bedroom.

Six months later I was promoted to Sergeant First Class and Ronald Reagan was elected President. I felt good about my career all over again. To get ready for the next higher rank I attended the Senior Non Commissioned Officer Course at division headquarters in Oklahoma City. I finished among the top students.

In the meantime Lieutenant Colonel Stoddard was reassigned and Lieutenant Colonel Charles L. McBride became the battalion commander. And then my friend Dick Smith applied for the next AGR position and came on active duty as a sergeant first class.

I had not forgotten my skills as a recruiter and in 1982 I recruited ten new soldiers for the 1st Battalion, 291st Regiment. The commanding officer of the Oklahoma City recruiting battalion noticed my actions and presented me his Commander's Certificate of Appreciation.

I was gratified to be recognized once again by the Army's recruiting command. However, it was the 3rd Brigade Commander, Colonel Dwight E. Whorton who gave me the first medal that included my efforts in recruiting.

In October of 1982 he awarded me the newly created Army Achievement Medal, recognizing, among other accomplishments, my efforts at recruiting for the 1st Battalion and the 3rd Brigade. This was the kind of recognition that eluded me while on active duty recruiting service helping to build the All Volunteer Army and bringing an end to the draft.

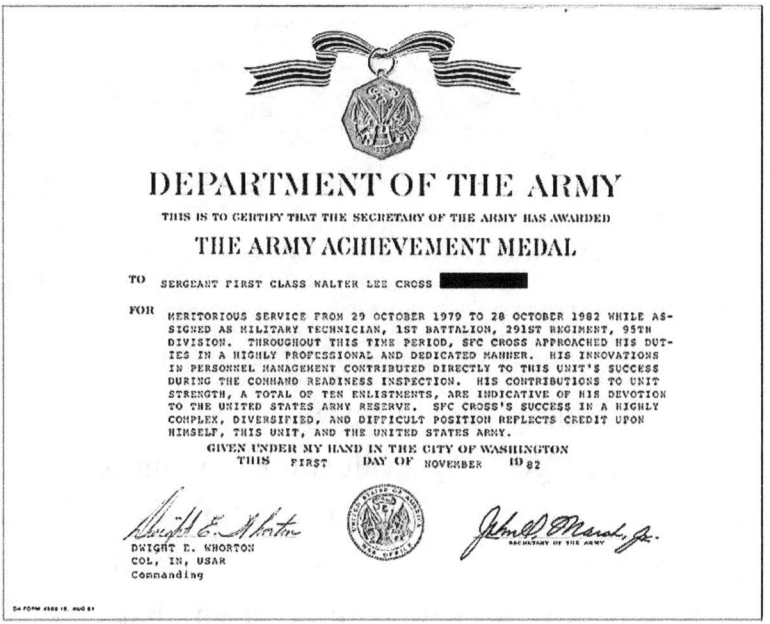

During this entire time period from October 1979 to October 1981 I received nothing but maximum scores on all my evaluation reports. And that was the way the remainder of my time in the 95th Division would continue. A year later I was awarded the First Oak Leaf Cluster to this same medal by Lieutenant Colonel Charles McBride for my accomplishments as the 1st Battalion's military unit technician.

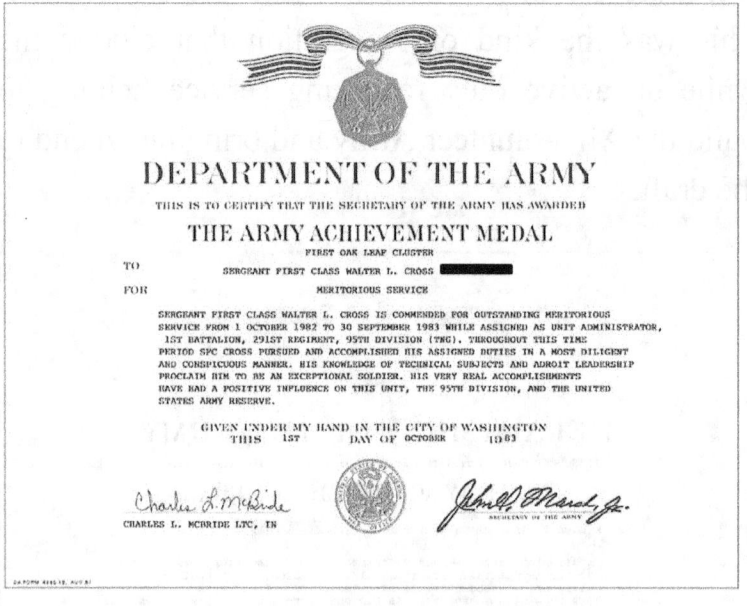

Positive things continued apace for the next two years and longer. My evaluations continued to be perfect and I received maximum rating scores in all areas. The upshot was that in May of 1985, just three years after my promotion to sergeant first class, I was promoted to master sergeant. Unknown to me I had also been recommended, selected, and placed on the short list for attendance at the residence course of the U.S. Army Sergeants Major Academy at Fort Bliss, Texas.

Selection to the academy is very prestigious and there are two ways to attend. One is through the non-residence course by correspondence, and the resident course at Fort Bliss. Attendance at the academy itself is the more prestigious of the two and I was on my way there.

At the time of my selection, and I assume even today, only the top four percent of the armed forces senior NCOs to include the Army, Navy, Air Force, Marines, and the Coast Guard, are selected to attend the U.S. Army Sergeants Major Academy. It was a singular honor for me to be chosen.

A rare photo of me in the old khaki uniform, 1983.

Reserve Center Named for Hero
by SFC Walter Cross

Sergeants First Class Richard L. Smith and Walter L. Cross flank the Cleo O. Payne Display at the Stillwater reserve center.
Stillwater News-Press Photo

The squad was returning to camp after a day-long search and destroy mission. The roads were frozen solid when they left that morning, perfect support for the heavy tanks of the 120th Combat Engineer Battalion. By the afternoon, the heat of the sun, and the churning tank treads had reduced them to a sea of mud.

Winding into a mountain pass, the column had nearly gained its protection when the rearguard tank lurched sideways off the road. Engine roaring, the tank's tracks flung chunks of mud rearward. The dozer blade on front of the tank was stuck deep in the bank along the road. Elevating the blade to its highest position did not free it. Tank Commander SSGT Cleo O. Payne dismounted, entrenching tool in hand, followed by all the crew except the driver. They dug at the glue-like mud. The sound of the engine drowned the distant thump of mortar shells leaving their launchers.

Payne and his men, working next to the stranded tank, were not aware of the Red Chinese barrage until the shells burst about them. The mud reduced the effectiveness of the exploding rounds, pelting the crew with harmless mud and water — all except one. SSGT Cleo O. Payne, of the 120th Combat Engineers, 45th Infantry Division (Oklahoma), gave the ultimate sacrifice. The date was February 16, 1952.

Thirty-one years later, in the Army Reserve center named in his memory, Payne's old uniform was discovered in a box tucked away in a forgotten corner. Sergeants First Class Walter L. Cross and Richard L. Smith, 1st Bn, 291st Regt, 3rd Bde, at Stillwater, OK, were determined to restore this symbol of the fallen warrior. With the help of SSGT Raymond Emde of the 45th Infantry Brigade, a personal friend of Payne, the unit crests, division patch, and rank for the old "Ike" jacket were obtained. Mrs. Carol Cross, a military wife for nearly 16 years, provided the sewing expertise. The uniform jacket is mounted in a display case at the reserve center, along with photographs and other memorabilia, for viewing by the public.

SFC Cross is the Retention NCO for 1st Bn, 291st Regt at Stillwater.

Above is an article I wrote for the Stillwater News-Press on Staff Sergeant Cleo Payne and the Army Reserve Center named for him, published in 1983. What follows below is the article as written that appeared in the News Press.

CLEO PAYNE DISPLAY: Local military historian, Sgt. First Class Walter Cross, right, and Sgt. First Class Richard Smith have constructed a display at the Cleo O. Payne Army Reserve Center here [Stillwater] honoring Payne, for whom the center is named. It involves Payne's recently-discovered uniform and a photo of Payne taken in his tank minutes before he was killed in action. (News-Press photo by Doris Dellinger).

His Uniform Discovered

By Doris Dellinger
News-Press Staff Writer

The date was Feb. 16, 1952. The country was Korea.

The squad was returning to camp after spending the day on a search and destroy mission according to local military historian Sgt. First Class Walter L. Cross.

When the men of the famed 45th (Oklahoma) Infantry Division left that morning, the roads were frozen solid, offering perfect support for the heavy tanks of the 120th Combat Engineer Battalion. But by afternoon, the heat of the sun and the churning tank treads had reduced the roads to a sea of mud.

"Winding into a mountain pass," Sgt. Cross says, selecting his words with precision, "the column had nearly gained its protection when the rearguard tank lurched sideways off the road. Engine roaring, the tanks' tracks gulped chunks of mud and flung them rearward."

The 'dozer blade on the front of the tank became stuck deep in the bank along the road. Elevating the blade to its highest position didn't free it.

The tank commander, Staff Sergeant Cleo O. Payne of Stillwater, dismounted, entrenching tool in hand. Followed by all the crew except the driver, he and his men began to dig at the glue-like Korean mud that held the tank.

"The sound of the engine drowned the distant thump of mortar shells leaping from their launchers," Cross points out, recreating the scene of 31 years ago.[48]

"Staff Sergeant Payne and his men, working next to the stranded tank, were not aware of the Red Chinese barrage from the low hills beyond them until the shells burst about them. The mud reduced the effectiveness of the exploding rounds, pelting most of the crew with harmless mud and water. All except one."

---

[48] As of 1983 when the article was written.

A shell [fragment] struck Sgt. Payne, killing him instantly. Last month, in the Army Reserve Center at 2715 N. Washington which is named for Payne, his old uniform was discovered in a box tucked away in a forgotten corner, Sgt. Cross explains.

Cross and Sgt. First Class Richard L. Smith, both of the 1st Battalion, 291st Regiment, 95th Division, say they decided to restore the uniform as 'a symbol of Stillwater's fallen warrior."

With the help of Staff Sergeant Raymond Emde of the 45th Infantry Brigade, a personal friend of Payne, the unit crests [insignia of the 120th Combat Engineers], the 45th Division's historic Thunderbird patch and Payne's rank for his old "Ike" jacket were obtained. Mrs. Carol Cross, a military wife for nearly 16 years, provided the sewing expertise.

The men mounted the uniform jacket beneath a draped American flag in a small wall display case in the entry of the reserve center. There the jacket, a Purple Heart Medal and other memorabilia are now available for viewing by the community.

There's even a photograph of the bogged down tank, affectionately nicknamed "Babe". Payne is shown emerging from the tank; a buddy stands alongside. The picture was snapped by one of the crew members just 10 minutes before Payne's death.

In the center of the display is a certificate of gratitude signed by President Harry Truman. It begins, "He stands in the unbroken line of patriots who have dared to die that freedom might live."

"Most people think the center is named for Payne County. We wanted them to know it was named in honor of Cleo O. Payne," Sgt. Smith, who share's Cross' interest in the preservation of military history, says.

The building is open Monday through Friday from 7:30 a.m. until 4 p.m. The public is welcome to stop by and see the display. [49]

Unit Crest 120th Combat Engineer Battalion

---

[49] Security requirements preclude easy access to the display. Contact the Cleo O. Payne Army Reserve Center commander for information.

.50 caliber machinegun training unknown major, Walt, unknown lieutenant, Captain (later major) Terry Allen on the drill floor of the Cleo O. Payne USAR Center, Stillwater, Oklahoma summer, 1985. Terry Allen was a good friend who left us much too soon. When I came back to the 291st Regiment after attending the Sergeants Major Academy, I worked directly with Major Allen.

CERTIFICATE OF PROMOTION
To all Who Shall See These Presents, Greeting:

Know Ye, that reposing special trust and confidence in the fidelity and abilities of WALTER LEE CROSS *I do promote him to* MASTER SERGEANT *in the* **United States Army** *to rank as such from the* twentyfourth *day of* March *nineteen hundred and* eighty five.

You will discharge carefully and diligently the duties of the grade to which promoted and uphold the traditions and standards of the Army.

Soldiers of lesser rank are required to obey your lawful orders. Accordingly you accept responsibility for their actions. As a noncommissioned officer you are charged to observe and follow the orders and directions given by superiors acting according to the laws, articles and rules governing the discipline of the Army, and to correct conditions detrimental to the readiness thereof. In so doing, you fulfill your greatest obligation as a leader and thereby confirm your status as a Noncommissioned Officer in the United States Army.

DONALD J. DELANDRO
Brigadier General, USA
Commanding

Promotion certificate to Master Sergeant

At this time the 3rd Brigade did not have a position for an AGR master sergeant and soon after I put on my new stripes, I got orders to proceed to the 5th Medical Group in Birmingham, Alabama. I was to serve in my old medical MOS of 91B50, senior medical NCO, as the group's operations sergeant.

This award should actually read FIRST OAK LEAF CLUSTER.

Once again I was awarded a decoration by the 3rd Brigade commander, at this time Colonel McEdward Law. The second Army Commendation Medal I received. This award covered my service from October of 1983 to August of 1985, a period of twenty two months. I didn't know it at the time, but Colonel Law was to be instrumental in my later return to Stillwater. But for now Carol, Justin, and I packed up and prepared to move to Birmingham, Alabama. We leased our house out and packed our bags. We found a beautiful home in the Birmingham suburb of Montevallo.

My son Justin aged five, and I on our balcony overlooking the back yard. Behind us is a national forest. Justin started first grade in Montevallo, and while we were there we took him to Disney World. We also got to watch a space shuttle launch from this same spot. I grew the beard while on leave. The picture is circa July, 1985 and was taken by Carol. We lived here only six months and then I was sent to the U.S. Army Sergeants Major Academy at Fort Bliss, Texas.

Chapter XIV
The 5th Medical Group (USAR)

A Recommendation to the Highest Enlisted Rank

After getting moved in, I drove the hectic morning commute to downtown Birmingham while Justin started first grade and Carol learned to maneuver through our three story house. The house backed up to a national forest and was complete with upstairs bedrooms a balcony and a full basement.

The only drawback to this house was the lawn, like all the other nearby houses, was home to several fire ant colonies. This was something I'd never encountered in Oklahoma. I learned to walk very quickly when I mowed over one of their nests!

I reported to the 5th Medical Group in Birmingham and met an extraordinary man and Army Officer, Captain Dallas W. Miller. We got along quite well, in fact more than quite well during the short six months I was with the 5th Medical Group. The reason for the short assignment was that I was officially alerted in my first month of duty at my new post that I was selected to attend the resident[50] course of the United States Army Sergeants Major Academy.

---

[50] There are two types of the USASMA course, the non-resident correspondence course and the resident course of six months. The

At the time only the top four percent of all senior NCOs from the five[51] branches of the armed forces were selected for attendance at the USASMA. I was to report in January of 1986 for Class XXVII. As I arrived at the 5th Medical Group in June of 1985, my time with the group and Captain Miller would only be six months.

Above is a picture of Captain Dallas Wadell Miller, the S-3 (operations) officer for the 5th Medical Group. He presented me this picture upon my departure from the unit and wrote a message on the back that reads:

"MSG Cros(s) A fine soldier and great individual. Good luck on your way to becoming

---

residence course is the more prestigious of the two.
[51] Including the U.S. Coast Guard.

CSM (command sergeant major) of the Army. Thanks a lot."

It will become clear later on why he said what he did about my becoming the top non-commissioned officer of the United States Army.

At the end of my tour in Alabama I was presented my third Army Commendation Medal and Dallas wrote me an efficiency report that not only gave me the highest of marks, but also recommended that I be considered for eventual selection, promotion, and assignment as the Sergeant Major of the Army.[52] The highest military assignment I was ever recommended for.

---

[52] There is only ONE sergeant major of the Army at any one time.

Captain Miller presenting me with my third Army Commendation Medal awarded to me by the 5th Medical Group's commander, Colonel Frank Young Jr. I never had the honor of meeting Colonel Young and so I asked that Dallas be the officer to present it to me. Below are some of my photographs from the 5th Medical Group.

On my right is Sergeant First Class Frank Trioli, assistant S-2[53] NCO, and to my left is Captain Miller. Both of these men had a great sense of humor and kept me laughing most of the time. As you can see I'm trying hard not to burst out laughing

---

[53] Security and intelligence section.

from something one or both of them just said. Trioli was a former special forces (Green Beret) NCO. You can faintly make out his airborne and combat infantryman badges above his name and U.S. Army patches.

The 5th Medical Group was a newly formed unit and while I was assigned there they got their new unit insignia. I volunteered to paint the new crest from the description given by the Army Department of Heraldry so they could display it. It was one of the 'extra' things I did that seemed to impress Colonel Young. These folks are from the S-2 Section including Sergeant First Class Trioli.

The officer to the left (a major) tried to convince Captain Miller not to recommend me for appointment as the Sergeant Major of the Army. But to no avail. I even suggested he reconsider writing that in my EER, but he insisted and persisted. His efforts were a singular honor to me.

Rank insignia of the Sergeant Major of the Army

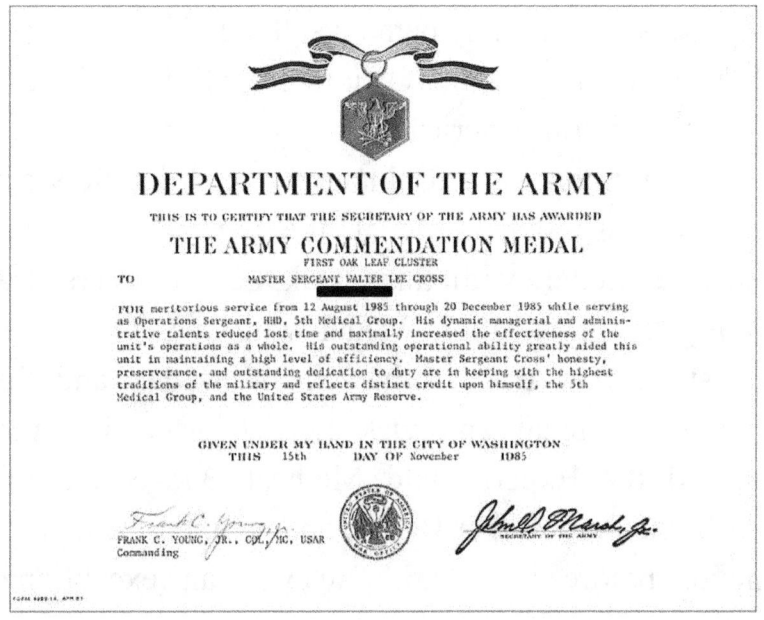

This award is from Colonel Young, commander of the 5th Medical Group and should actually read 'SECOND OAK LEAF CLUSTER'.

After I left the 5th Medical Group for Fort Bliss, Texas Captain Miller was, at his request, reassigned to the same post with the 3rd Cavalry's 'Brave Rifles'.

He came to a couple of the academy's 'dining in' ceremonies as my guest. He also came to my graduation; he was a good man, a good officer, and a good friend. I last saw him in the summer of 1986. He retired from the Army and died much too soon in February of 2009. We had served together during two of our assignments and after Fort Bliss, Dallas was stationed in Georgia, Berlin Germany,

Honduras, and in Panama as the Public Health / Environmental Health Officer for all US Military Forces in Latin America.

His military awards include: the Joint Services Meritorious Service Medal, the Army Meritorious Service Medal, with three Oak Leaf Clusters, the Army Commendation Medal with two Oak Leaf Clusters, the Army Achievement Medal, and the Overseas Medal 2nd Oak Leaf Cluster. He also earned the Expert Field Medical Badge and the Airborne Paratrooper Badge, and attained the rank of major before his death. He was an exceptional officer and I will always remember him.

Coincidentally he later served in a Buffalo Soldier reenactment group as the first sergeant while I served in the Sons of Union Veterans as a Union Army Sergeant Major. Also we were similar in our service to veterans groups. What a great guy and outstanding Army officer!

Carol was very active in the lady's group known as the Ultima Wives.[54] One of the events the ladies put on was skits of life at the academy. Many of the women got all dolled up and put their best foot forward like a beauty pageant. But Carol dressed up as a master sergeant in parts of my uniform, wore a

---

[54] The group's name is derived from academy's motto *Ultima* (ultimate).

bald rubber cap, put a pillow under the shirt and a cigar in her mouth. She pantomimed as a student attending the academy. Her antics and clever phrases brought the house down in laughter and congratulations for the best costume. It was the soldiers that gave her a standing ovation. She loved it!

While I was in attendance at the Sergeants Major Academy there occurred one of the worst disasters in the history of our space exploration program. I am referring to the destruction of the space shuttle *Challenger*. Approximately one month after my arrival at the academy, on January 28, 1986 the shuttle lifted off. Aboard were seven highly trained astronauts.

When I was assigned the task of helping create the class book for Class XXVII, USASMA I along with others recommended we dedicate the book to the crew of the *Challenger*. We did so, and the photograph displayed below appeared at the front of the book.

The astronauts aboard that fateful flight were: (starting back row, left to right) El Onizuka, S. Christa McAuliffe, Greg Jarvis, Judy Resnik, Mike Smith (pilot), Dick Scobee (commander), and Ron Mcnair.

Photograph courtesy of NASA.

Chapter XV
Return to the 1st Battalion

Graduation from the U.S. Army Sergeants Major Academy

In June of 1986 I graduated from the U.S. Army Sergeants Major Academy and I needed a duty assignment. I was told that my assignment would likely be to the Military Personnel Center in St. Louis, Missouri. However, a day or so later I got a call from Colonel McEdward Law, commander of the 3rd Brigade of the 95th Division back in Stillwater. He asked if I wanted to return to the 3rd Brigade and I told him I most certainly did. I soon found myself on the way back for my second assignment to the 291st Infantry Regiment, USAR. The only difference was I would be with a different battalion other than the 1st Battalion. Boy, it really turned out to be a big difference!

When a soldier receives more than one of any decoration, the second and subsequent awards are indicated on the medal's ribbon with an oak leaf cluster (usually). Since this was my second assignment to the 291st Infantry Regiment after serving in two other organizations, I think I should have been able to wear on oak leaf cluster on my unit crests. So, I invented one as you can see in the graphic above.

The 3rd Brigade and the 1st Battalion were in very poor shape administratively when I arrived. It was surprising to me how fouled up the two units had become in one year. In my absence my friend Dick Smith was promoted to master sergeant and was reassigned away from the 1st Battalion. He had already been gone for a number of months.

The battalion's pay records were all FUBAR[55] resulting in the low morale of the soldiers and creating a general overall dissatisfaction. In addition the brigade's training records administration had suffered and was likewise very fouled up. The cherry on top was that the 3rd Brigade had an Inspector General Readiness Inspection scheduled in the very near future.

---

[55] Fouled Up Beyond All Recognition.

Although assigned to a battalion outside Stillwater I found my work week split between Stillwater and the battalion in another town about sixty miles away, and the 1st Battalion's many problems.

From the moment I reported to the commander of this other battalion he made it very clear that I had been forced upon him and his command. He had already picked out the man he wanted to be his AGR operations sergeant and it wasn't me. From that time on it became a real struggle between myself and this Army Reserve lieutenant colonel.

I am not going to mention this man's name but I will say he was an alcoholic and a lawyer. Good gravy, what a miserable combination! He was, bar none, the most incompetent, lazy, and inept officer I had met besides the adjutant I mentioned earlier when I was serving in Germany.

This lieutenant colonel did everything he could to make my assignment to his unit as miserable as possible. The first crack out of the box he wrote me a substandard efficiency report because he felt I neglected his unit while working on the 1st Battalion and 3rd Brigades many problems. However, he was so inept that he wrote it for a time period less than the required minimum time period of 90 days. I easily challenged it and had it removed from my personnel records. When he was informed that his evaluation of me was invalidated he was livid!

I should mention here that this battalion had two other AGR soldiers serving with it, both sergeants first class. In addition to my full time duties to all three units, I served as the operations NCO for the battalion, and the first sergeant of its Headquarters Company.

In the meantime, I got the payroll of the 1st Battalion straightened out and was instrumental in getting an AGR sergeant first class assigned as that unit's new unit administrator.

That man was Sergeant First Class Billy Ray Steptoe. And although he came to resent me for the role I played in getting him assigned to the battalion, he did an exemplary job, as I knew he would. But he wasn't real happy about leaving his native state of Georgia.

I helped the 3rd Brigade get their needed 'Satisfactory' rating on their Readiness Inspection. And then I did something to alleviate my own situation with the problem lieutenant colonel.

The 3rd Brigade was now under the command of Colonel Dwight E. Stoddard, my old commander from the 1st Battalion. I wrote him a letter and told him what was happening to me at the hands of the battalion commander. I was honest with him and he responded immediately! He made the 3rd Brigade in

Stillwater my duty location and my last year on active duty I did my very best for both him and the 3rd Brigade.

My branch told me that my graduation from the sergeant's major academy almost certainly guaranteed my promotion to sergeant major. But when I learned my next assignment would be to the Military Personnel Center in St. Louis I turned the opportunity down and retired. I'd pretty well had all I wanted although I knew I would miss both all of my comrades and the challenges of continuing to serve.

I retired as an AGR soldier with more than twenty years of active duty and began to receive my retired pay at age 38. I was reassigned to the Retired Reserve until May of 1997 when I received my final honorable discharge from the U.S. Army.

Some years after I retired one of the men from the 1st Battalion, retired sergeant first class Jerry Ward, reached his 62nd birthday. He came by my house with his paperwork to apply for his Army retired pay. I told him it was his task to fill out the forms and send them in. His reply was

.50 caliber machinegun training in the Stillwater USAR training center. Yes, that's my class ring from the Sergeants Major Academy.

"But Walt, you're my unit administrator, and a lot of the guys have had a lot of trouble getting paid."

I laughed I had not been a unit administrator for at least twenty years or more.

"Okay Jerry, I'll take care of it for you." I said, and I had him sign the forms. Happily he never had any trouble getting paid. So I guess he did right.

Not long after I retired the 291st Regiment and the 3rd Training Brigade were deactivated. We still meet however in an annual reunion each October, have dinner, and as old soldiers are wont to do, tell a lot of old war stories. A good friend of mine, Chief Warrant Officer Five Norman Filtz (U.S. Army Reserve Retired) was and is the organizer of the reunion. I thank him for the many, many, memories!

He has held the old unit together for a lot of years. We will have our eleventh reunion in October of this year (2018). May they all live long and be happy.

95th NOTES

No, that's not Don Prudomme and the Army's double "A" fueler, but it is Justin Cross and if you look close you'll see the Army Reserve decal on his pinewood racer. Justin is the son of Master Sgt. Walter L. Cross of the 4th Battalion, 3rd Brigade. He recently raced his car at the Cub Scout's pinewood derby and won! He's got a blue ribbon to prove it.

PROMOTIONS
4073d US Army Recpt Bn

To Specialist 4
Eaton, Susan E
Tynes, Rebecca L

To Sergeant
Dargin, Christopher T

THE UNITED STATES OF AMERICA

TO ALL WHO SHALL SEE THESE PRESENTS, GREETING: THIS IS TO CERTIFY THAT THE PRESIDENT OF THE UNITED STATES OF AMERICA AUTHORIZED BY EXECUTIVE ORDER, 16 JANUARY 1969 HAS AWARDED

THE MERITORIOUS SERVICE MEDAL

TO MASTER SERGEANT WALTER L. CROSS, HEADQUARTERS, 3RD BRIGADE
95TH DIVISION (TRAINING), STILLWATER, OKLAHOMA

FOR EXCEPTIONALLY MERITORIOUS SERVICE WHILE SERVING AS THE MILITARY TECHNICIAN (AGR) FOR THE 1ST BATTALION AND THE 4TH BATTALION, 291ST REGIMENT, AND LATER AS THE SENIOR OPERATIONS NON-COMMISSIONED OFFICER FOR THE 3RD BRIGADE, 95TH INFANTRY DIVISION (TRAINING) FROM 1 OCTOBER 1979 TO 30 JUNE 1988 INCLUSIVE. THROUGHOUT THIS TIME PERIOD MASTER SERGEANT CROSS DEMONSTRATED AN EXCEPTIONAL GRASP OF ART OF THE SENIOR NON-COMMISSIONED OFFICER BY HIS OUTSTANDING LEADERSHIP, ACCEPTANCE OF AN EXPANSION OF HIS RESPONSIBILITIES, AND THE DEDICATED DETERMINATION TO INSURE THE SUCCESS OF THE TWO BATTALIONS AND THE 3RD BBRIGADE AS U.S. ARMY RESERVE COMMANDS.

GIVEN UNDER MY HAND IN THE CITY OF WASHINGTON THIS 30TH DAY OF JUNE 19 1988

Dwight Stoddard
DWIGHT STODDARD
COL, IN
COMMANDING

The award of the Meritorious Service Medal from Colonel Stoddard would be the third decoration awarded to me by a 3rd Brigade commander, and the final decoration for my military career. I think it was appropriate that it would be from Dwight Stoddard, an exceptional soldier and unit commander and a good friend! My family and I are most grateful to him.

DEPARTMENT OF THE ARMY
HEADQUARTERS, 95TH DIVISION (TRAINING)
POST OFFICE BOX 10095
MIDWEST CITY, OK 73140-1095

PERMANENT ORDERS 28-1   17 April 1988

CROSS, WALTER LEE,   , MSG, Headquarters, 3d Brigade (OSUT),
95th Division (Training), Stillwater, Oklahoma

Announcement is made of the following award.

Award: The Meritorious Service Medal
Date(s) or period of service: 29 October 1979 to 1 July 1988
Authority: AR 672-5-1, para 2-17
Reason: For outstanding meritorious service
Format: 320

FOR THE COMMANDER:

MICHAEL H. BUCKLEY
MAJ, AG
Assistant G-1

DISTRIBUTION:

1 - CDR, 95TH DIV (TNG)
1 - IND CONCERNED
1 - IND MPRJ
1 - 227-03 FILE
1 - 227-16 FILE
1 - CDR, FIFTH U.S. ARMY
    ATTN: AFKB-PR-RPD-RE
1 - CDR, ARPERCEN
    ATTN: DARP-EPS-A

Employer Support of the Guard and Reserve

In May of 2007, partly because I had served in both the Regular Army and the U.S. Army Reserve, I was selected as the Executive Director of the Oklahoma Committee of the ESGR by the Chairman of the state committee a retired Army National Guard brigadier general.

The Employer Support of the Guard and Reserve is a Department of Defense program that insures returning veterans of the Army and Air National Guard and the Army, Air Force, Navy, and Marine Reserve are protected in the work place upon their release from full time active duty. I was hired by the Department of Defense and worked out of an Army National Guard base in Oklahoma City for three years.

During my tenure as executive director two employers from Oklahoma, one of which was the Choctaw Tribe, received the Freedom Award from the Department of Defense. I traveled twice to Washington D.C. to represent the state at these two prestigious awards.

When I retired in November of 2009 after three years as the committee's executive director, I had finished my service to the nation both as a soldier and a Department of Defense government employee.

At the end of my service to the ESGR I received a citation, embossed by the Great Seal of the Choctaw Nation of Oklahoma. It reads as follows;

Dear Mr. Cross;

I wanted to say thank you to you and the Oklahoma Employer Support of the Guard and Reserve for not only awarding me with the Patriotic Employer Award, but also for the opportunity [to compete for the Freedom Award]. I do believe that our soldiers should be supported and recognized for their diligence and dedication. We should be supporting and uplifting our troops as they do our great country.

Again, thank you for your recognition. Sincerely,

Gregory E. Pyle, Chief
Choctaw Nation of Oklahoma.

This citation hangs proudly in my home.

Appendix – A Vietnam Photo Album

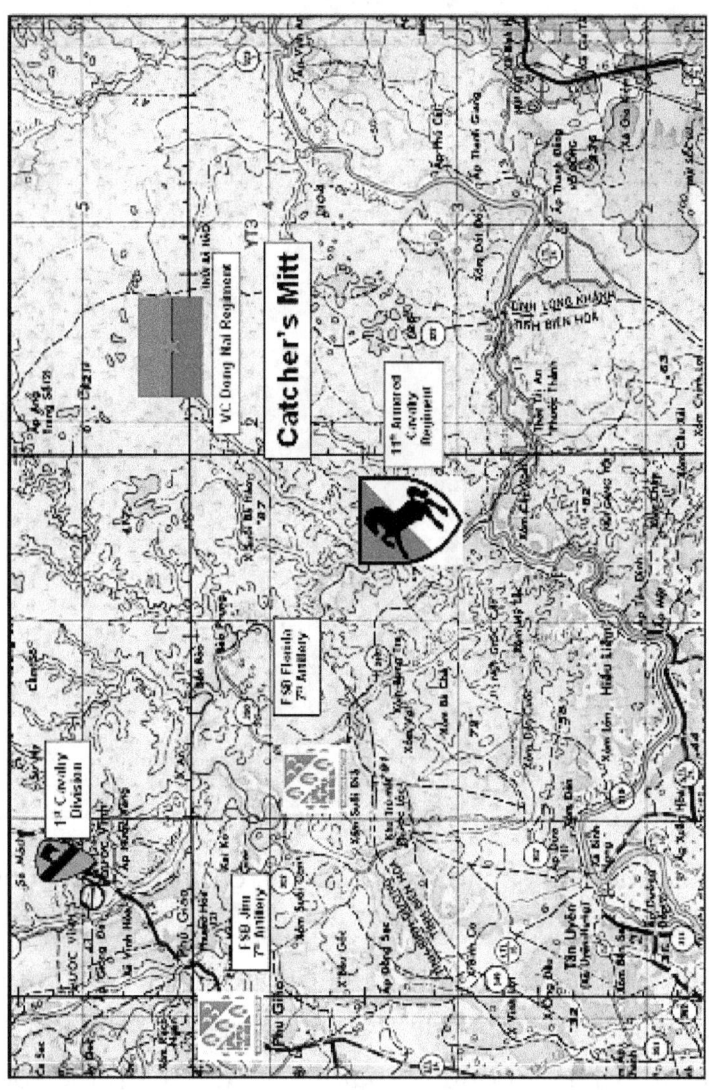

War Zone D showing the location of FSB Jim and FSB Florida.

I included this photo album to remind the reader I began my service in Vietnam War. These first two photographs are from Fire Support Base Huertgen near the Iron Triangle. Redlegs rarely wore shirts. I was all of 135 pounds soaking wet! This rather poor quality picture was taken with a Polaroid camera.

This is a good illustration of the older type of howitzer (M-101) with its sandbag parapet surrounding the position. These cannon were later replaced with the M-102. The flag in the background is the unit guidon.

Captain John Dubia, Battery A commander and his executive officer at Fire Support Base Huertgen. The battery's guidon showing its designation as A, 1/7th is clearly seen. Captain Dubia eventually attained the rank of Lieutenant General (3 stars). I understand that A Battery was his only combat command in a long and illustrious Army career.

Not readily visible in this photo is the fact I had a "new guy" sunburn. This is the first photograph of me in the field. It's Miller time!

I kept my driving skills sharp by driving the battery commander's jeep inside Fire Support Base Aachen. We were now in the Iron Triangle, an area that had always been under the control of the enemy.

I and a fellow redleg stand inside a gun position termed a 'parapet' at Fire Support Base Aachen. The guard tower is clearly visible in the background.

Note the tip of artillery rounds or "projos" (short for projectiles), behind us. This photo was taken in March of 1969. This is a well-constructed gun position.

Artillerymen are nicknamed 'Redlegs' a name given them by the infantry because of the uniform they wore during the American Revolution that consisted of a blue coat and red trousers.

Every soldier I ever knew has taken one of these "Machine Gun Kelley" kinds of photos. The tent right behind me was my home, known as a 'hooch'.

You can see the jeep I was in for the earlier picture over my right shoulder. The tree line behind me was about 300 meters away, which gives you an idea of the size of the rubber trees surrounding Fire Support Base Aachen in the Iron Triangle.

Another photograph similar in nature at Fire Support Base Aachen in March 1969.

By this date Richard M. Nixon was President of the United States.

The note on the back of this photo, taken and sent home to my parents, says this soldier is my best friend. I've forgotten his name but his nickname was "the killer" (for killing rats). We are moving by convoy to Fire Support Base Jim, April 1969. My hand rests on a box containing loaded M-16s.

A scene from Fire Support Base Jim; a fire mission is in progress, note the dust rising from the concussion of the gun's firing and the gun is back in recoil. The man on the left is our chief of the firing battery, known as the 'chief of smoke'. His rank is sergeant first class.

An early scene at Fire Support Base Jim taken in May of 1969 outside the Exec Post[56], the battery guidon flutters above my head. This kind of head gear was called a 'go to hell' hat by the soldiers but was more correctly termed a 'slouch' hat.

---

[56] The executive post was where the executive officer was stationed and served as the battery's orderly room and headquarters.

In this bull session, the sergeant standing in the middle of the photo, like myself, was recommended for the award of the Silver Star for "Gallantry in Action". He also earned the Purple Heart. The oldest man in this photo is 21 years of age. The kid sitting down was a very cocky individual. But then I guess we all were. It was kind of a defensive mechanism.

Ready for action are artillerymen John Jordan, Thomas Bonk, and Paul Jones. Paul and I correspond to this day and he still refers to me as "Doc".

Doc's office: that is a patient in my 'waiting room' with my medical supplies behind him and my M16 above him on the wall. The walls were made of dirt filled artillery shell ammunition boxes and covered in sand bags.

I slept in this bunker along with the battery commander's jeep driver.

August 31, 1969 at Fire Support Base Jim. The tent behind me was my "hooch". I'm wearing unauthorized airborne combat boots that zip up on the inside. Note the friendship bracelet given to me by Babysan, the daughter of our laundress, on my right wrist. It was not long after this photograph that the battle for Fire Support Base Jim took place on September 9, 1969.

The water puddle behind us indicates it just rained. First Sergeant Jerry Cooper sports a band aid from a small wound received during the fight for Fire Support Base Jim, September 1969. The supply sergeant to my left brought us steaks and beers from the base camp at Phuoc Vinh. Both men received Bronze Star Medals for their service in Vietnam.

Phuoc Vinh was the First Cavalry Division base north of our position. On a trip to that base is where I last saw and spoke to David Lemon a friend of mine from my high school days in Perry, Oklahoma.

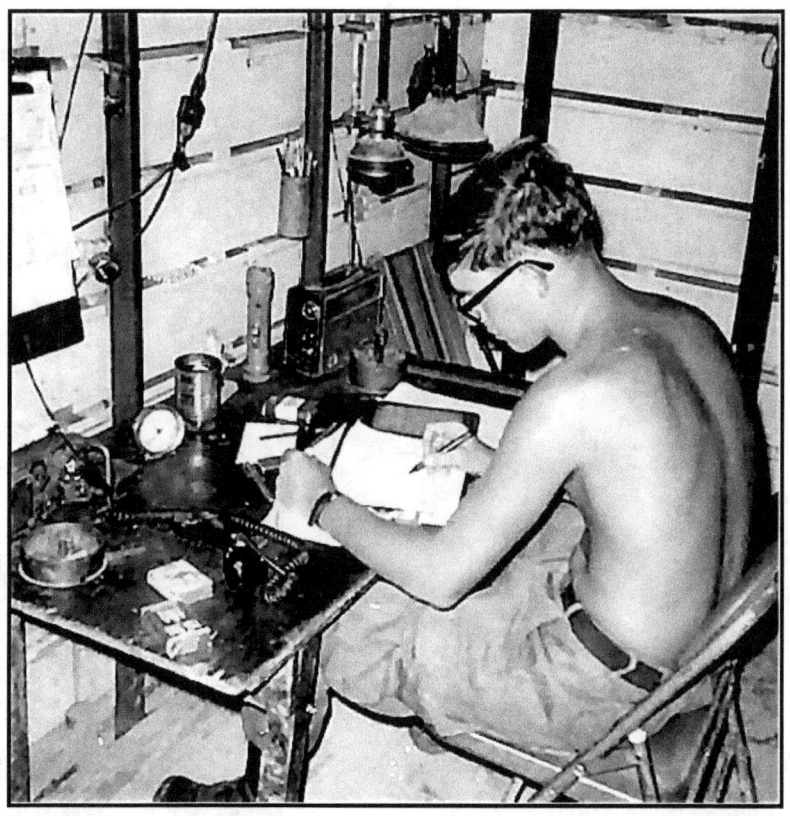

Here I am in the Exec Post recording a fire mission. The ashtray was made from the base of an expended artillery round. I have an engraved one to this day presented to me by Major Bob Church a staff officer of the battalion as a memento of my service in the 7th Artillery. The radio handset lies on the field desk.

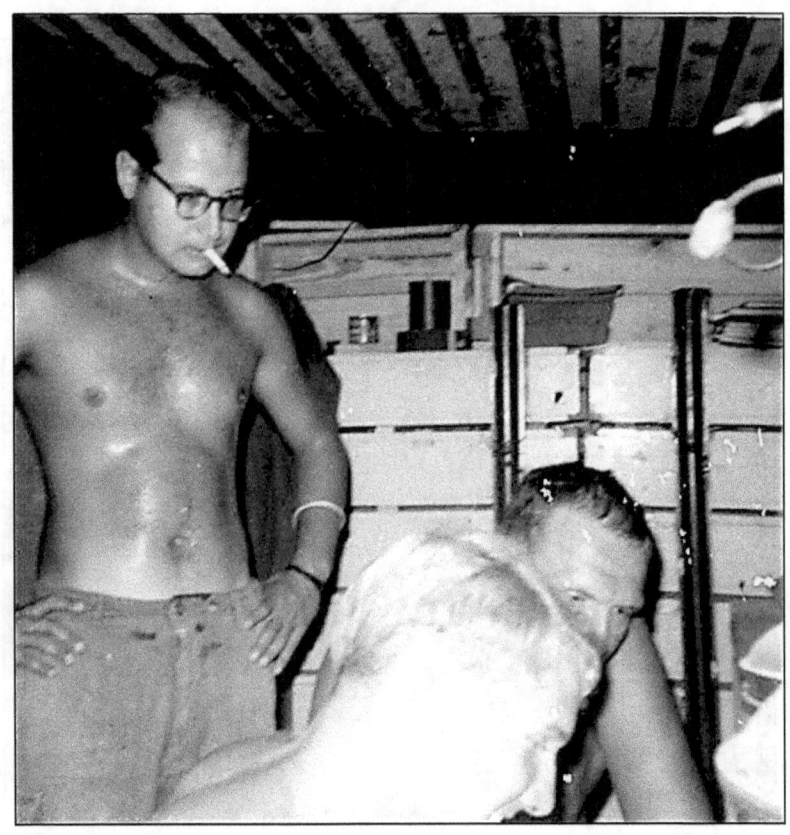

Exec Post members Bob Poteet standing and John Nagley sitting with First Sergeant Jerry Cooper. Poteet was known for his seemingly endless grumpiness. Nagley was the exact opposite, always smiling, as he is in this photograph.

A shot of the shower stall about to be put to good use. This shower stall was destroyed twice by direct hits from enemy mortar rounds. I always suspected it was the target they initially aimed for and then would adjust their fire.

In this photograph I am decorated by Colonel Sperow at Fire Support Base Jim, it was an exceptionally beautiful and cool day in Vietnam. I had just been promoted to Specialist 5 (same pay grade as a sergeant) November '69. I received a Bronze Star Medal with 'V' (valor) device for heroism in ground combat. I received a second award of the Bronze Star Medal in December just before I left Vietnam to return to 'the real world'. After the awards presentation, we had a luau complete with a roast pig in a pit. Colonel Sperow served all us our plates of ham and all the fixings.

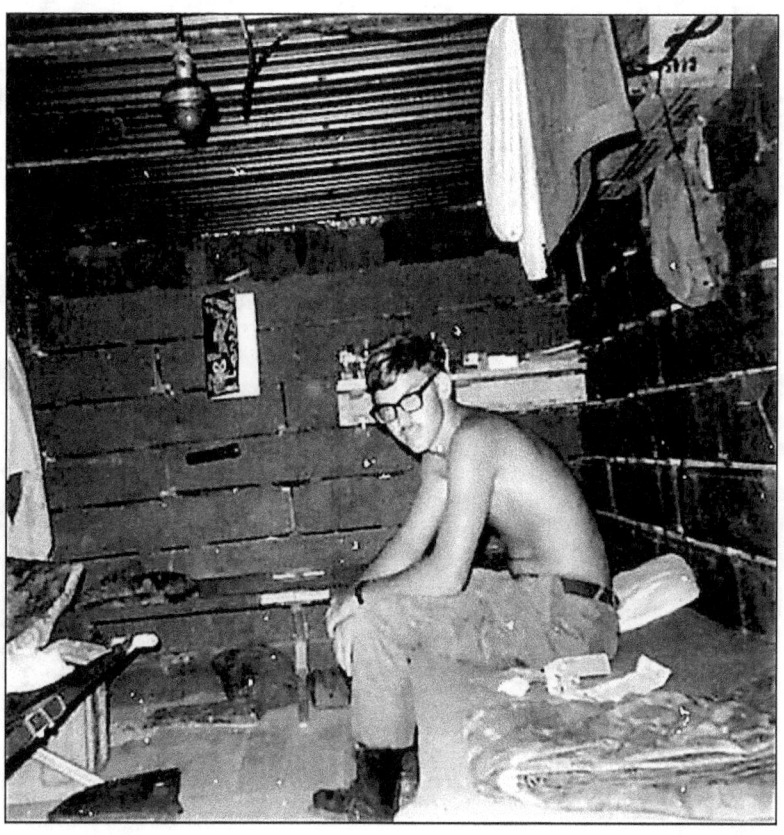

My bunk inside the 'hooch' I shared with two other soldiers.

In this photograph 'Smiley' Nagley, my nickname for him, is impersonating the battery commander, Captain Harry G. Madden

The real Alpha Battery Commander, Captain Harry G. Madden both an excellent Army officer and a superior battery commander.

The average age of the men of Alpha battery was twenty years old. Captain Madden wasn't much older, perhaps his late twenties, but despite his age, he was the "old man". It was Captain Madden, with concurrence of the battery first sergeant that recommended me for the Silver Star Medal. We sometimes called him the "BC".

Behind the captain you can see a stream of water from the monsoon runoff. A heavy gauge steel construction piece known as PSP forms a foot bridge for easy passage. PSP was used throughout Vietnam from aircraft runways to the floors and roofs of bunkers. It was one of the most commonly used types of construction material besides steel engineer stakes and sandbags.

A hot day somewhere in Vietnam in November of 1969. Note the friendship bracelet is gone. I mailed it to Carol for her little sister Rhonda.

This is the last photo taken of me at Fire Support Base Jim and indeed in Vietnam. I'm standing in front of my medical supply trailer. Soon after this picture we headed out to Fire Support Base Florida, in December of 1969. I was approaching my DEROS day.[57]

---

[57] Date Eligible to Return from Overseas.

Base plate of a 105mm round fired by the 7th Artillery early in the war, sometime in 1965 or 1966 courtesy of Bob Church. Note the date of "1945" indicating this round was created while my dad, a WWII veteran, was still in the Army. This "trench art" ashtray sat on Bob's field desk in Vietnam.

This is some of the engraving on the ash tray given to me by Bob Church. Also on the sides is the unit DUI and "1st Battalion 7th Artillery" all filled in with artillery red. It is one of the prized possessions from my military career.

In February of 1968 Specialist Russell Gallegos arrived in country and was assigned as an artilleryman to Battery A, 1st Battalion, 7th Artillery. He recalls his arrival in his own words.

"When we arrived in Vietnam (aboard a commercial airliner) we could not land because the airfield was under mortar attack. We detoured to the Bien Hoa Air Base where we would have ended up anyway.

There, overnight, we got in a deuce and a half (2 and ½ ton truck) headed for Di An. Upon arrival I was issued an M-14, web gear, flak vest, and assigned to a gun crew. I changed into jungle fatigues and put my personal effects in a locker. Staff Sergeant Growler, my new crew chief, collected me and my first detail was to fuse a load of 105mm rounds and store them in a bunker. There was a blonde headed corporal I later learned was from Ohio named Edward Koehn, he eye balled me and said 'Get your ass in here and get to work Poncho Villa' the nickname stuck, Ed Koehn and I remain friends and close as brothers to this day. I'd do anything for him, and although he lives 1500 miles from me, we keep in touch."

The battery supported elements of the 1st Infantry Division who secured the massive Tan Son Nhut Air Base during the communist offensive of Tet 1968.

Each day of the offensive the battery's round output doubled from the day before.

The following month Major General Keith L. Ware too command of the division as the Big Red One took part in Operation Quyet Thang (Resolve to Win). One month later and the division was involved in the largest operation of the Vietnam War, Operation Toan Thang (Certain Victory). A scant six months after taking command, on 13 September 1968 General Ware was killed in action when his command helicopter was shot down by enemy anti-aircraft fire. Battery A, and other elements of the battalion fired a continuous fire mission until the general's remains and those of his crew and staff were recovered.

The division assistant commander, Major General Orwin C. Talbott assumed command of the division.

In the meantime Russell found himself outside the village of Cu Chi supporting the 25th Infantry Division and serving as the radio operator for a forward observer. In an old Buddhist temple he met a medic named Joaquin, also from Texas. They took a liking to one another and together sang Mexican songs in the evening. Joaquin made fun of Russell because he could not speak Mexican Spanish.

While on patrol Russell was shocked to discover a dead ARVN soldier hanging by his neck. No one knew if the VC had done it or maybe other

Vietnamese soldiers. They also found an enemy weapons cache and Russ helped himself to an AK-47 that he carried throughout the remainder of his duty tour.

When Russ returned to the battery's firing position he helped fire a mission in support of the very outfit he had been on patrol with. During a close combat firefight Joaquin was hit and killed in action. Russ was saddened by the death of his new friend. In his honor Russ later named his son Christopher Joaquin Gallegos.

The battery rotated back to Di An and then on to a new position in the Iron Triangle. But not before enemy sappers came on the base throwing satchel charges in the barracks. The battery fired illumination and set up to use killer junior[58], but the enemy just melted away.

Russ found the assistant gunner had rotated home and he now took that position. The gun chief told Russ he wanted their gun to be the fastest gun in the battalion. He had Russ work become closely familiar with the hook, rope, and wooden handle of the artillery gun lanyard. Once he had its operation down the gun was firing between fifteen and eighteen shells per minute. That was fast!

---

[58] An artillery round that fired 20,000 flechettes (small metal darts), this shell was also known as a 'beehive' round because of its distinctive sound.

During the TET offensive the battery was constantly moving to support elements of the 25th Infantry Division, the 1st Cavalry Division and our own Big Red Unit units. Sometimes they would move forward in support of the infantry and then back to the base in the same day. When the battery moved the mess hall went with them, and Russ states they only ate combat rations once.

During our long stay at Fire Support Base Jim the mess hall did come out for a while. The head cook, a sergeant first class, worked at night doing the baking during the cool hours. Part of my job as the medic was to insure the sanitation of the mess hall. So sometime during the night, usually when I could smell the pastries baking, I inspected the mess tent. And of course I had to sample those pies that were cooking and make sure they were hot and delicious enough, especially those cherry pies! They always were.

Russ and the battery pulled into a new position near the Cambodian border and in late June of '68. The first order of business was to build a gun parapet. It took three days and nights to fill the fifteen hundred sandbags to make a good fighting position, doing the work between hot fire missions. And just as the parapet was completed the battery was moved to An Loc.

It wasn't long before an emergency fire mission came down. An infantry company was in danger of being overrun by a strong Viet Cong force. The infantry called for fire on their own position and although we received no official word on casualties either friendly or enemy we got a thank you for getting the VC off their ass. It was hard to sleep for the next few nights.

"Back in the Iron Triangle and near the Black Virgin Mountain (Nui Ba Den) we moved in support of the Rome plows. The battery was airlifted into positions to protect the engineers operating the plows which were stripping the jungle away. The VC was not happy about being exposed to our aircraft and ground forces. With the removal of trees and brush the spiders, centipedes, scorpions and snakes were looking for new homes, with us GIs. We followed the plows for two weeks until we came to a road. There the battery loaded up on trucks for a road movement. While moving ammunition I and two other battery members were stung by scorpions. The medics treated us with aspirin, water, and rest and we climbed up inside the truck and went to sleep, a little unexpected in country R&R.

During the monsoons I managed to get a bunk space in a tent and moved out of the rat infested bunker I had been staying in. About eight o'clock that evening I heard a distinct thud and I knew the rain had caved a bunker in. I rushed and found it was the bunker Marvin Schwint was sleeping in. I started pulling sand bags off not sure which end of the bunker he was in, and I called for more help. I had got lucky and guessed correctly and after removing only five or six sand bags I found him. He gasped for air as I pulled him up and I noted a wound on his forehead where the steel reinforcement of the ceiling had hit him. He was air lifted to a field hospital and was okay. When he came back he told me I was the prettiest thing he had ever seen when I pulled that sand bag off him."

Sergeant Russ (Poncho) Gallegos Battery A,
1st Battalion, 7th Artillery (Pheons)

"I am glad to have served with some real American soldiers."

The photographs that follow are from Russ.

Specialist (later sergeant) Russell Gallegos and Sergeant Jim Frisk at Lai Khe, Russell provided the following important photos after contacting me in September 2012. Our tours of duty only overlapped by a month (Jan – Feb 1969), I am grateful to him for providing them.

In a ceremony in September 1968 Major General Albert E. Milloy came to Battery A, 1st Battalion, 7th U.S. Artillery in Lai Khe to fire the one millionth round sent down range by the battalion. Preparing for his arrival is a four man gun crew. The first two soldiers from the left are Sergeant Jim Frisk and another soldier identified only as Moon. Sergeant Frisk's gun was selected to fire the millionth round because his gun was the oldest howitzer in the 1st Infantry Division.

General Milloy (far right back to camera) arrives for the ceremony accompanied by battalion commander Lieutenant Colonel Merle Crocker (first full figure from left back to camera). The man just to the left of the general is likely the battery commander. The soldiers from left to right (center) are Sergeant Jim Frisk, Moon, Robert Chappa and an unidentified soldier.

General Milloy pulls the lanyard and fires the one millionth 1st Battalion, 7th Artillery round to be fired in Vietnam. There would be another quarter million rounds fired before the battalion left for the U.S. The Oklahoma and California state flags fly in the background.

Left to right: Staff Sergeant Growler, Bob Reed, Russ Gallegos, Ron Mallard, Bob Pollack, Ed Koehn, unknown soldier in February 1969, Di An.

Russ and other gun bunnies distribute artillery shells to the separate gun sections. This appears to be an Iron Triangle position, likely FSB Huertgen.

Russ' caption for this picture reads "Ed Koehn and Doc, Iron Triangle". No, that's not me. This is the medic I replaced in January 1969 when he was either wounded or injured. Everyone knew him (and later me) as "Doc" and no veteran of the 1/7th I've contacted remembers his actual name.

Left to right is Corporal Ed Koehn, unknown, and Ron Mallard assembling projectiles at FSB Huertgen, January 1969. The last two men look surprisingly alike.

Following the Rome plows in the Iron Triangle near Nui Ba Den (Black Virgin Mountain, 996 meters high of solid rock with a thin veneer of soil), the battery was airlifted in to support jungle clearing operations. In this photograph a Chinook helicopter brings in one of the battery's 105mm howitzer, its ammunition conveniently slung underneath. This mission lasted two weeks. Russ and others got stung by scorpions here.

Bob Pollack poses with an enemy rifle captured during the Rome plow operation. It is not the famous AK-47 which is an automatic rifle. It is a bolt action rifle, possibly a rifle left over from the Second World War. In some instances these kinds of weapons could be taken home as souvenirs, but full automatic weapons like the AK were forbidden.

Fire mission! This looks like a Thunder Road position, that's Russ with his face an inch or two from the sight, Bob Mallard standing with his back to the camera, Bob Reed in the background. The other two men are unknown. A spent round lies on the ground with others on the ready board awaiting the designated powder charge (1 to 7) and type of fuse, usually, fuse quick. Note the sand bags atop the ammo bunkers.

End of fire mission. Left to right unknown; Bob Reed, Russ, Bob Mallard, unknown.

Joey Bishop visits the battery in November of 1968. Russ says he was a nice guy, humbled that he was among fighting soldiers. Bob Pollack (glasses, looking away from camera, Bob Mallard in background, Staff Sergeant Growler, Joey Bishop with Russ standing directly across from him (back to camera), Staff Sergeant Downing, and Corporal Ed Koehn. I never heard of this visit by Joey Bishop mentioned, and I arrived only about a month later.

Iron Triangle, January 1969. That appears to be a command Loach helicopter flying into the battery area. It's quite likely the battalion commander (by this time Lieutenant Colonel Francis King) visiting the battery which had just come under the command of Captain John Dubia. And around the time the battery received a new "Doc". Specialist Walt Cross.

Sometimes the engineers were slow removing trees that could provide a hiding place for the enemy. A quick 'direct fire' mission could remove a tree quite handily, and provide a little target practice for the gunners.

The last picture Russ took in Vietnam of the 1/7th Artillery.

This is my friend Tom Bullock who served as a member of the 377th Security Police Squadron at an air base near Saigon in 1969. Tom was in the Air Force at that time. While at FSB Florida, I was only about 20 kilometers from his position. When the base came under enemy probing or attack, we [Alpha Battery, 1/7th] provided supporting fire for Tom's unit and the rest of the air base. It would be nearly forty years before I would meet Tom and exchange war experiences discovering this link between us. I met Tom when I went to Washington D.C. for ESGR (Employer Support of the Guard and Reserve) training in 2007.

Here I am at the Vietnam Veterans Memorial Washington, D.C. in September 2008. The day I visited it was very hot and humid, appropriate Vietnam weather. In May of 2010 six more names were added to the wall. These were veterans of the war who sustained wounds so severe that they eventually succumbed to them years later. They are Army Lt. Col. William Taylor, Marine Corps Lance Corporals John Granville and Clayton Hough Jr., Marine Corps Corporal Ronald Vivona, Army Captain Edward Miles and Army Sergeant Michael Morehouse, comrades all.

The artillery adds a bit of class to what would otherwise be a vulgar brawl. This is the Distinguished Unit Insignia (DUI) of the 7th Artillery. The motto "Nunquam Fractum" is Latin for "Never Broken". The three arrowheads are called "pheons" in heraldic terms and that was the nickname of the men who served with the 7th Artillery in Vietnam. It was a name that would remain in Vietnam, because when the unit returned stateside in March of 1970 it was redesignated the "7th Field Artillery" and the new insignia, assigned by the U.S. Army Institute of Heraldry, did not include the pheons.

NUNQUAM FRACTUM – NEVER BROKEN

Above is the he 7th Artillery Vietnam pocket patch. Normally, unit insignia is made of metal and worn on the shoulder strap of the dress uniform. In Vietnam it was cloth and worn on the left pocket of the combat fatigue uniform shirt. The colors are silver and red. Red is the official color of the United States Army Artillery. The Army hymn was initially the official song of the artillery. St. Barbara is the patron saint of the artillery.

Master Sergeant Walter L. Cross, 1986.

This is the official photograph of me taken at Fort Bliss, Texas at the U.S. Army Sergeants Major Academy. My promotion packet for sergeant major went forward in February 1988. However, the family decided it would be better for us all if I retired rather than be transferred away from our Stillwater, Oklahoma home. Oddly, once I had decided to retire, it didn't seem like it had been twenty years plus since I first enlisted. I would, indeed, miss the Army.

These are my personal and unit decorations. The bottom right decoration is a commemorative medal recognizing my participation in the Cold War. This decoration is a commercially created medal because despite the importance of winning the Cold War, the Department of Defense has never approved an official decoration.

About the Author Walt Cross

I had three goals for my life's work that I decided early on in my life. My first and earliest goal was to be a soldier. This goal was inspired by my dad's service in World War II. I succeeded in that particular goal.

My second goal was to be a park ranger. I completed that goal in 1989 one year after my retirement from the Army. On April 1, 1989 just a couple of weeks from my fortieth birthday, I went to work for Oklahoma State University as the ranger for Lake Carl Blackwell. Lake Carl Blackwell is a recreational park located about nine miles west of Stillwater and the OSU campus.

My job was to supervise the recreating public on the land and waters of the park. To that end I drove a patrol pickup, and operated a patrol boat on the lake surface. I was the ranger of Lake Carl Blackwell for four years while I earned my master's degree from OSU. It was a great experience and I learned a good deal about the interaction of people in a recreational environment. After graduation in the spring of 1993 I left this adventure exactly four years to the day after starting it. Afterward I continued employment with OSU until I retired in 2006.

The third goal I had set for my life was to be a writer. I started writing in earnest in 2003, and just weeks after retiring from the university I completed writing and published my first book. Here is the title of that book and of other books I have written.

The Frontier Army of the Indian Wars

Custer's Lost Officer; the Search for Lieutenant Henry Moore Harrington, 7th U.S. Cavalry

From Little Big Horn to the Potomac; the Story of Army Surgeon Dr. Robert Wilson Shufeldt

Out West with Custer and Crook; the Story of Colonel Verling K. Hart of the Seventh and Fifth U.S. Cavalry

A Strange Look at Custer
(Written in conjunction with author/illustrator Jack Strange and featuring his exceptional illustrations)

The U.S. Army in World War II

From the Beaches to the Baltic; the Story of the 7th Armored Division in WW II
 (Contributing editor to the unit's compiled history)

Red Tracers; the 482nd Anti-Aircraft Artillery Battalion in WWII

Flash – Bang; the Unit History of the 285th Field Artillery (Observation) Battalion

Vietnam War

The Fastest Gun in the Big Red One; the 7th U.S. Artillery in Vietnam 1965 – 1970

Fighting the Second Vietnam War at Oklahoma State University

Science and Ancient History

Did God Kill the King of Sodom?

Fiction

Prodigal Moon; the Arrival of the Ninth Planet

The Legacy of Ivanhoe

I have edited and published many World War II unit histories annotated for clarity for a 21st century reading public. These books keep the heroes of WWII in the public view, and every one of them is dedicated to my father.

These books are all published by Dire Wolf Books and available online at:
www.lulu.com/greenpheon7.

www.ingramcontent.com/pod-product-compliance
Lightning Source LLC
LaVergne TN
LVHW022111080426
835511LV00007B/750